CHERNOBYL

THE NATURAL ENVIRONMENT —
Problems and Management Series

Edited by *Chris Park, Department of Geography,*
University of Lancaster

CHERNOBYL

THE LONG SHADOW

By Chris C. Park

ROUTLEDGE
London and New York

First published 1989
by Routledge
11 New Fetter Lane, London EC4P 4EE
29 West 35th Street, New York, NY 10001

Printed and bound in Great Britain by
Biddles Ltd, Guildford and King's Lynn

British Library Cataloguing in Publication Data

Park, Chris C.(Chris Charles), *1951–*
    Chernobyl: the long shadow. — (The Natural
    environment, problems and management series)
    1. Ukraine, Chernobyl — Nuclear power stations.
    Accidents, 1986
    I. Title II. Series
    363.1'79

    ISBN 0-415-03553-8

Library of Congress Cataloging in Publication Data

Park, Chris C.
    Chernobyl.

    (Routledge natural environment — problems and
management series)
    Bibliography: p.
    Includes index.
    1. Radioactive pollution — Environmental aspects —
Europe. 2. Chernobyl Nuclear Accident, Chernobyl,
Ukraine, 1986 — Environmental aspects. 3. Nuclear
industry — Public opinion. I. Title. II. Series.
TD196.R3P37 1989   363.1'79   88-32353
ISBN 0-415-03553-8

CONTENTS

TABLES

FIGURES

Dedicated to the memory of

Ivy Hannah Park (née Connell)
1892 - 1960

ACKNOWLEDGEMENTS

My ability to write this book owes much to others to whom I am greatly thankful. I am endebted to the University of Lancaster for granting me the study leave which gave me time to write most of it, and to my colleagues in the Department of Geography for not disturbing me (too much)! The book has been produced camera-ready copy, using Aldus Pagemaker software on an Apple Macintosh micro and Laserprinter from files initially created in VIEW on a BBC Model B micro. I have written, edited and set up the whole text; it has been quite an experience! I owe thanks to Claire Jarvis and Elsa Drinkall for teaching me how to use the Apple Mac. Claire Jarvis, valued friend and trusty cartographer, drew the maps and diagrams (often from amazingly scrappy originals) with customary flair, and was a source of great encouragement throughout.

At Routledge I am grateful to Eleanor Rivers for guidance and support and, as always, to Peter Sowden for his faith in me.

My parents, Alex and Margaret Park, as ever in the background, were there when I needed them. I am blessed with a wonderful circle of good friends at St Thomas' Church in Lancaster who continue to encourage me. I particularly value the love and support of Claire Jarvis (again!), Peter John and Ann Davies, Robin and Jill Bundy, Philip and Gill Gower, and Stephen and Audrey Potter.

Life would not be the same without Angela, my cherished companion, critic, best friend and wife. She tolerates my long spells glued to the word processor with remarkable patience; she knows when to remain silent; she is there when I need support and love. Our son Samuel was born during Chapter Eight. He has brought us great joy and is priceless; my next book will be written for and dedicated to him. This book is dedicated to the memory of my grandmother, who died far too young. I have nice childhood memories of her; I wish we could have had her around longer.

PREFACE

The world will never be quite the same again after Chernobyl. The accident at the nuclear power station on Saturday 26 April 1986 - by far the worst nuclear accident anywhere yet - spilled out vast quantities of deadly radioactive material which was blown by the wind over many thousands of square kilometres. Thirty people died within a month or so from radiation poisoning, all of them workmen or firefighters at the station on the Saturday.

Chernobyl cast a long shadow over the whole of Europe. Fallout from the accident was recorded in every country in western Europe. Radio-active dust contaminated milk, killed reindeer, made roast lamb an unpopular Sunday dinner. The shadow will also last a long time. Whilst no-one outside the plant died directly, over a hundred thousand people living within a 30 km zone were exposed to dangerously high radiation levels before they were evacuated. Many are likely to contract cancer in the next thirty years, and many of these will die as a result of Chernobyl. Further downwind, especially outside the Soviet Union, people were exposed to levels of radiation which were much lower but still well above background levels. Some doctors estimate that thousands of Europeans could die from cancer caused ultimately by Chernobyl.

Within six months of the accident, which blew open number four reactor at Chernobyl, numbers one and two reactors had been switched on again and were producing much-needed electricity for the heavy industry in the Ukraine. But the area around the station was still heavily contaminated. Local towns remained deserted. It was a ghost landscape, in which the only signs of life were troops and workmen cleaning down buildings and roads, and abandoned dogs roaming freely in search of food.

The long shadow over western Russia left the country's traditional "bread-basket" heavily contaminated with deadly radiation. It will be many years before people can safely return to their homes and farms, and before the fields can be safely used to grow crops again.

The final toll, in human misery and hardship as much as in dead and injured, will be high. This was the accident nobody thought would happen. Yet it did, and the world watched from a distance. Outside the Soviet Union politicians quickly concluded that such an accident could never happen in their country - nuclear reactor designs are different (and better), safety procedures are better worked out and easier to implement, operators are better trained. A thousand reasons were given.

But these assurances did little to comfort the many millions of people around the world who watched daily news reports about Chernobyl on television. Somewhere, out there, an invisible cloud of radiation was floating about freely overhead, roaming wherever the wind systems cared to blow it. Rains were washing the deadly radiation down to earth in some places. All people could do was watch and hope. The shadow was long!

Disaster are rare, thankfully. Major disasters, whose effects ignore national boundaries and spill over into neighbouring countries, are exceedingly rare. When one happens there is immediate interest in how people cope - the human loss and suffering, the sacrifice and bravery, the recriminations over who was to blame. The media are good to spot a good story. After all, bad news is good news for the media. When hard information is thin on the ground, as is typically the case when major disasters strike, speculation, inference and pure fabrication get mixed up with fact. Sensationalism and drama go hand in hand to make good copy.

But the benefit of hindsight is a great thing, because it reveals the lasting legacy of all disasters - the lessons to be learned. After the surge of humanitarian interest subsides, and the sensation-seeking media drop the story in pursuit of more topical items, life goes on for those directly and indirectly affected by the disaster. Their fame and the broad base of public interest is short lived. But there is an obligation - borne out of respect for those who died and were injured, and for those who will face problems (health, economic or other) in the future, and out of concern to minimize the risk of repeat performances elsewhere (insofar as this is possible) - to carefully study the lessons and explore what they imply.

This book tries to offer a retrospective on what happened during and after the Chernobyl accident, and a prospective on the lessons for the future. At the time of the accident, I was working on a book on acid rain, widely regarded as *the* environmental problem of the 1980s. The long shadow of Chernobyl shares some significant properties with acid rain. Both are invisible forms of air pollution. Neither respects national or administrative boundaries. Both have attracted widespread public and political interest around the world. Both pose serious threats to health. Both produce long-term problems over a wide area, which will ultimately prove to be much more serious (and costly) than the immediate problems. Both have been created by the energy industry. Ultimately, both stem from an unbridled use of technology and a fundamental lack of concern about the long-term health of mankind and stability of the environment.

1
ALARM BELLS RING

Before the end of April 1986, Chernobyl meant nothing to the average person. But the accident at the Chernobyl nuclear power station quickly projected it into the forefront of public attention; it held a prominent position on television news bulletins and was on the front page of leading newspapers around the world for many weeks. Now the very name Chernobyl is taken as synonymous with nuclear accident. Rightly so, many argue, because it was by far the worst accident ever to occur in a nuclear power station. In terms of damage done and anxiety caused, if not immediate death toll, it also rates (alongside Bhopal) as perhaps the worst technological disaster of all times.

The accident clearly deserves full analysis and this can only be done some time after the event as the initial speculations, so eagerly and speedily broadcast through the news media, have been superceded by more sober factual accounts. The results of detailed studies of what happened, how it was coped with, and what it implies in the long term started to appear within six months, and it is now possible to reflect on Chernobyl in context.

In this book we shall look at what happened, how it affected people both locally and further away, and what lessons can be drawn from the unfortunate affair. But we should start at the beginning, when the first signs of anything untoward landed - quite literally - out of the blue over Sweden, over a thousand kilometres away.

THE STORY BREAKS

At 2.00 pm on Sunday 27 April 1986 automatic radiation measuring instruments at the then unmanned Swedish National Defence Research Institute (FDA) in Stockholm registered a marked rise in radiation levels in the air over the city. The unexpected rise must have caused alarm when it was discovered by staff who turned up for work the next morning. The cause was a puzzle to Swedish scientists.

There were only two possible explanations (other than instrument malfunction), and neither was good news. It could be fallout from a nuclear weapon and - whilst it was hoped that one had not been unleashed with serious intent anywhere - there were suspicions that the monitors might have picked up evidence of atmospheric nuclear testing believed to be under way in China. The alternative was accidental leakage of radiation from a civil nuclear installation, probably a nuclear power station. Either way, it was clear that something significant had happened, somewhere, recently.

Meanwhile, the Monday morning shift was finishing at the Forsmark nuclear power station on the Baltic coast, about 100 km north of Stockholm (Figure 1.1). At 9.30 am radiation detectors 4 km from the station picked up high levels of radioactivity, and unexpectedly high levels were measured on the shift workers' clothing as they passed through the routine monitors before going home. Checks on radiation levels on soil and vegetation around the site confirmed that radiation levels were much higher than normal, at around 0.1 millisievert an hour (units are explained in Table 1.1).

Figure 1.1 The unfolding drama - location of places mentioned in the text

The obvious explanation was an undetected radiation leak on site. Operators at Forsmark have three alert conditions, depending on the level of radiation which is found. Below 0.025 millisievert an hour is *blue alert*, between 0.025 and 1 millisievert an hour is *yellow alert*; above 1 millisievert an hour is *red alert* (the most serious). The station was put immediately on *yellow alert*; the reactor was closed down, all 600 workers on site were evacuated, and the search was on for the leak. Although radiation levels were between four and five times higher than normal background levels, they were not high enough to justify evacuation of the population around the power station.

Within hours, by early afternoon on Monday 28 April, further pieces of the jigsaw became available. Monitoring stations throughout Scandinavia were now recording abnormally high levels of radiation in the atmosphere - up to six times the norm in Finland, five times the norm in Sweden and Denmark and up by half in Norway. The Swedish National Institute of Radiation Protection (SSI) announced that it had registered radiation just about everywhere in the country it looked. The Swedish Defence Ministry was quick to point out that although levels were higher than normal there was no danger to people in Scandinavia.

The hunt is on

Swedish scientists carried out detailed laboratory tests to discover exactly what was in the cocktail of radioactive material falling over the country. They found three main ingredients - traces of the noble gases xenon and krypton, fairly high levels of iodine and caesium, and heavy elements like neptunium. This chemical finger-printing allowed them to rule out the prospect of fallout from a nuclear weapon and confirmed that somewhere the core of a nuclear reactor had been badly damaged. The question was, where?

Neither Denmark nor Norway has any nuclear power stations, so they looked anxiously to their Nordic neighbours for information on the likely source. Sweden had never experienced any accidental nuclear releases of its own but its sensitive radiation detection systems, which are amongst the most sophisticated in the world, had in the past measured minute levels of radiation which had leaked from Sellafield (formerly Windscale) in Cumbria, north-west England - nearly 1,500 km to the south west (Figure 1.1).

Weather patterns over Europe at the time suggested that for several days the wind had been blowing northwards from Russia, up from the Black Sea around the Ukraine and over the Baltic into Scandinavia. Three likely sources of the fallout were identified, each a major nuclear power

Table 1.1 Summary of radiation terms

It is difficult to make sense of many reports about radiation, because the units used are unfamiliar to most of us. The problem is made even worse by the fact that not all reports use the same units, because of recent changes in the types of measurements used. 'New' International System (SI) terms (*becquerels, sieverts, grays)* have replaced 'old' terms (*rems, rads, curies*).

Radiation terms express three inter-related but different things:

(a) ACTIVITY : this expresses amount of radiation (more precisely, the amount of radioactivity in a substance), and it reflects the speed at which a radioactive element spontaneously decays (or disintegrates) and releases its energy.

Becquerel (Bq) : the new term, corresponding to the decay of one atom per second. It is much smaller than the old term *curie* which it replaced.
Curie (C) : the old measure of rate of decay, equal to 37 billion (37,000,000,000) becquerels. A *nanocurie* is one billionth of a curie, equal to 37 becquerels; a *millicurie* is one thousandth of a curie, equal to 37,000,000 becquerels.

These terms are often used to describe how much of a radioactive substance (eg caesium-137) is measured in air, on the ground or in food. In unit terms, these are expressed as becquerels per kilogram (Bq kg^{-1}) (eg in vegetables), per litre (Bq l^{-1}) (eg in milk), per square metre (Bq m^{-2}) (eg on the ground) or per cubic metre (Bq m^{-3}) (eg in air).

(b) ABSORBED DOSE : this expresses the dose of radiation absorbed by a body or substance, normally in terms of energy transfer (eg joules per kilogram).

Gray (Gy) : the most recent unit for measuring the amount of radiation which produces a certain electric charge (measured in joules) in a kilogram of dry air, equal to 100 rad.
Rad : a similar but older measure, equal to a hundredth of a gray (1 Gy = 100 rad). Equivalent to an energy absorption per unit mass of 0.01 joule per kilogram of irradiated material (ie 100 ergs per gram)
Roentgen : the amount of radiation which produces 0.258 coulombs of electric charge in dry air (NOTE that the coulomb has now been replaced largely by the joule). A *microroentgen* is a millionth of a roentgen.

(c) DOSE EQUIVALENT : this reflects the biological significance (eg to human health) of an absorbed dose of radiation, expressing potential harm regardless of the source or type of radiation.

Rem : a measure of the probable absorption of radiation by a human (regardless of the type of radiation involved - alpha, beta or gamma rays), short for *roentgen-equivalent-man,* being the amount of radiation which

Table 1.1 (continued)

will produce the same biological effects as one roentgen per kilogram.This unit of radiation dose is calculated by multiplying the dose (in rads) by the Relative Biological Efficiency of the type of radiation giving the dose (see Chapter Two) (ie rem = rad x RBE). This unit replaced the roentgen, and in turn has been replaced by the sievert (1 rem = 0.01 Sv)(but is still widely used amongst scientists). A *millirem* (*mrem*) is a thousandth of a rem (1,000 mrem = 1 rem).

Sievert *(Sv)* : new unit to replace rem, being a hundred times greater (1Sv = 100 rem). A *millisievert* (*mSv*) is a thousandth of a sievert (1,000 mSv = 1 Sv); a *microsievert* (*µSv*) is a millionth of a sievert (1,000,000 µSv = 1 Sv).

SOURCE: based on Salo (1986), Loprieno (1986), Hohenemser *et al* (1986), Wilkie (1986b)

station in western Russia (Figure 1.1). Leningrad (about 700 km east of Stockholm) could be ruled out straight away, because it did not fit the wind patterns. The next suspect was the world's largest operating nuclear reactor at Ignalina in Lithuania (around 700 km south-east of Stockholm). The nuclear power station at Chernobyl in the Ukraine could not be ruled out, although it was further away (some 400 km further in the same direction). The puzzle persisted ... where was it coming from?

Diplomatic fallout

It was clearly by now no longer simply a scientific problem, the politicians had to be informed. Neither was it any longer a national problem for Sweden, it was quite obviously a major international issue which might have untold consequences. Given the massive distances the radioactive cloud must already have travelled, radiation levels close to the source must have been dangerously high, many people were likely to have been exposed to lethal doses of radiactivity, and vast areas must already have been contaminated before the cloud was detected in Scandinavia. Western observers immediately saw the prospect of this being the world's worst nuclear accident, and the most serious blow yet to the peaceful use of atomic energy.

Diplomatic wheels were set in motion and early on Monday afternoon Sweden, through its Embassy in Moscow, asked the Soviet Union for information on what had happened. The initial response by the Soviet

Atomic Energy Authorities was to deny knowledge of any nuclear accident in the Soviet Union (although it was later to emerge - and be confirmed by the Soviet authorities - that an accident had happened, at Chernobyl, at 1.23 am on Saturday 26 April 1986 (see Chapter Seven)).

A day is a long time in politics, and through Monday representations were made to the Soviet Union by a number of countries seeking information. Sweden took the lead. Mrs Birgitta Dahl (Energy Minister) called upon the Russians to inform the rest of the world of nuclear accidents - especially if radiation was likely to spread to other countries - in good time and to improve their nuclear security. Hers was not the only call for all Russian nuclear reactors to be placed under international control. Norway's Environment Minister (Mrs Rakel Surlein) announced the likelihood that representatives would soon be sent to Moscow to lodge a formal complaint about the "regretful" incident at Chernobyl.

Bowing to international pressure, the Soviet Council of Ministers issued a brief statement through the news agency *Tass*, which was broadcast on Soviet television at 9.00 pm that evening. It was a fairly bald statement which gave a number of facts:

(1) an accident had occurred at the nuclear power plant at
    Chernobyl, north of Kiev in the Ukraine
(2) the accident had damaged the atomic reactor
(3) there were some casualties
(4) measures were being taken "to eliminate the consequences
    of the accident" at the plant
(5) aid was being given to those affected by the leak
(6) a government commission (Kremlin inquiry) was being set
    up immediately to investigate the cause of the accident

The statement was unprecedented in three ways. It was issued with a speed best described as uncharacteristic - once television news reports of the increased radiation levels were being broadcast around the world, the Soviet authorities realized they could no longer cover up the event. Secondly, the admission that there had been a major technological accident was without precedent. Disasters are hardly ever reported within the Soviet Union (or at least they were before the recent wave of *glasnost*), and when they are strenuous efforts are made to play down details of damage, casualties and death tolls. Thirdly, the Soviet Union maintains tight controls over 'strategic' information, including details of its nuclear programme (which is shrouded by a largely impenetrable veil of secrecy). The revelation that something had gone wrong in its nuclear power programme was an unprecedented concession which highlit the obvious seriousness of the accident.

There was a general feeling of anxiety in many countries over the way in which knowledge of the accident had emerged, not from the Soviet Government but via monitoring in Scandinavia. This immediately brought into question the integrity and judgement of the Soviet leadership, which was to be closely monitored over the following months (see Chapter Ten). Many, including Britain's Margaret Thatcher (speaking in the House of Commons on Tuesday 29 April), believed that all countries which belong to the International Atomic Energy Authority (IAEA), and this includes the Soviet Union, had a duty to report all nuclear accidents without delay.

IAEA, ever cautious, clearly wished to avoid the prospect of being drawn into the diplomatic exchange to dismiss any implication that it lacked confidence in Soviet abilities to cope. It quickly announced from its base in Vienna on 29 April that "although members are not bound to report accidents to us, we expect to be informed in due course. We are in the meantime trying to get what information we can, and we have a team on hand to send to Russia if required. However, the Soviet Union is well equipped with experts, and frankly we do not expect to be asked".

NUCLEAR POWER STATIONS

To understand what happened at Chernobyl, and why, we need to know something about how a nuclear power station works and what nuclear processes are involved.

The basic structure of a nuclear power station is not all that different from a more conventional fossil-fuel power station (Figure 1.2). In both, water is heated and turned into steam. The steam is then passed through a series of turbine blades, to rotate them. This in turn rotates a shaft in the adjacent generator which produces electricity, which can be fed via transformers into the supply system. The steam is then condensed back into water, and recycled.

The fission process

The furnace and boiler in the conventional power station are replaced by the reactor in the nuclear power station as the heat source for steam generation. The heat is generated within the reactor core, by the controlled release of radiation from fuel rods. The process is called fission, and it is

7

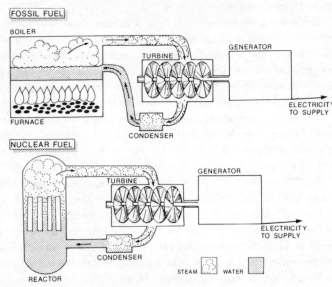

Figure 1.2   Basic structure of fossil fuel and nuclear fuel electricity generating systems

based quite literally on *"splitting the atom"*.

An *atom* is the smallest unit of matter (so small that ten million might fit side by side on a full-stop on this page). Whilst an atom is - by definition - indivisible, it can in fact be split if it is hit by a fast moving particle called a *neutron.* The basic fuel source in an atomic reactor is uranium. Three things happen when an atom of uranium is split:

   (a) the nuclear reaction releases a large amount of energy
   (b) the original relatively heavy atom is divided into two
         lighter ones (called *'fission fragments'*)
   (c) several new neutrons are formed

If the newly formed neutrons are allowed to collide with more uranium atoms, they in turn will be split (releasing energy, liberating neutrons and creating lighter uranium atoms). And so on. A self-sustaining *chain-reaction* (Figure 1.3) is triggered off, which is capable of producing vast amounts of energy from fission of relatively small amounts of uranium. Hence nuclear power, at least in theory, is an extremely efficient means of producing electricity.

Controlling the chain-reaction

But the chain-reaction must be carefully controlled. If there is enough uranium present and the process is not controlled, the fission process will automatically continue to accelerate and soon be impossible to control. Uncontrolled fission reactions provide the basis of nuclear weapons (which use plutonium as a fuel) and are capable of causing large-scale devastation, as we know from the bombings of Hiroshima and Nagasaki in Japan in 1945.

The heart of a nuclear power station is the *reactor core*. Small solid pellets of the fuel (normally powdered uranium dioxide) are packed inside *fuel pins* (narrow tubes up to 3.7 m long). Up to 200 fuel pins are then mounted together side by side into a *fuel element* (*or assembly*), which is placed inside the reactor core. The core will normally house a large number of fuel elements.

The fission process is controlled in two ways (Figure 1.4). The core is packed with a moderator material which absorbs (or moderates) the fast neutrons; this means that it slows them down and prevents uncontrolled fission reactions. Graphite, similar to the graphite used in lead

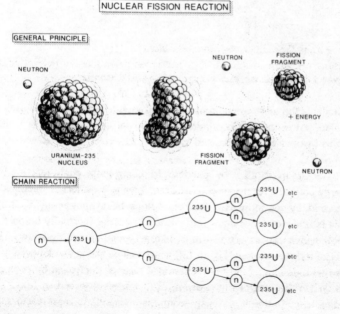

Figure 1.3   The nuclear fission chain-reaction   (after Miller 1979)

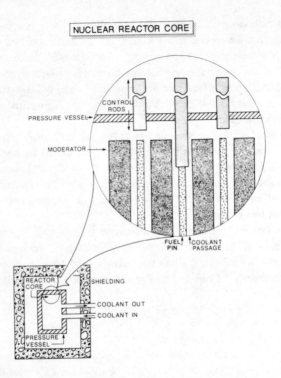

Figure 1.4 The nuclear reactor core and its main components

pencils, is the most common moderator. Graphite moderators are used in nuclear power stations in Britain and the Soviet Union, whilst the United States favours water moderated systems (Pressurized Water Reactors, or PWRs).

But it must also be possible to control the chain-reaction (thus energy output) over the short term, even down to a matter of fractions of a second, by controlling the number of atoms being split at any one time. This is achieved by control rods made of material (normally boron and steel) which can absorb neutrons, like a sponge absorbing water. The control rods are sleeves (Figure 1.4) which can be raised and lowered over the fuel rods in a carefully orchestrated manner, controlled by computers. When the control rod is down, neutrons are absorbed so fewer are available for fission (thus energy output is reduced). In most reactors the control rods can be dropped down instantly (under their own weight) to stop the chain-reaction; this effectively shuts down the reactor.

The reactor core generates immense heat, so it has to be cooled continuously. Commercial reactors in Russia are cooled by water which is circulated through coolant passages within the reactor core (Figure 1.4). Most British nuclear power stations are cooled by gas (they are called Advanced Gas Reactors, or AGRs).

The core is normally encased in a strong pressure vessel, which in turn is surrounded by protective shielding to prevent leakage of radioactive fission products into the surrounding environment.

THE CHERNOBYL NUCLEAR COMPLEX

People in the west knew little about Chernobyl before 28 April 1986. Most cared even less. Yet this was the jewel in the crown of the Ukrainian electricity supply industry, hailed as ultra-safe and highly efficient.

Chernobyl lies over 600 km south-west of Moscow (the Russian capital) and about 130 km north-east of Kiev (capital of the Ukraine and one of Russia's oldest cities). It is located beside a large lake on the Pripet Marshes, close to the confluence of the Pripyat and Uzh rivers (which flow through a reservoir into the Dnepr - Europe's third largest river - just north of Kiev). This is a farming area, and the power station brought life and resources, both of which were warmly welcomed. The new city of Pripyat had been built 14 km away, to house the plant's workers and their families. Nearly 49,000 people were living there by early 1986, and many thousands more were living in nearby Chernobyl town.

Chernobyl is more properly referred to as a nuclear complex than a power station. It has four large water-cooled, graphite-moderated nuclear reactors, each one capable of producing 1,000 megawatts of electric power. Two more identical reactors were under construction in April 1986. Building work began in 1971, and the first reactor went critical in August 1977. By December 1983 all four were on stream.

RBMK reactor design

The reactors were of the RBMK design, which is unique to the Soviet Union. They use fuel pins of uranium dioxide (slightly enriched in uranium-235) encased in a zirconium alloy cladding. Eighteen fuel pins form one fuel assembly, two of which are placed end-to-end in the pressure tube (fuel channel) through which cooling water is circulated. A total of 1,661 fuel assemblies are set vertically in a core containing around 200 tons of graphite moderator and 211 electrically-driven boron-carbide control rods. The reactor core, at 7 metres high and nearly 12 metres in diameter, is much larger than its US (PWR) equivalent.

RBMK stations are designed as twin-reactor stations, which means that two reactor cores are built side-by-side, sharing a refuelling machine and other auxiliary equipment.

The RBMK reactor is known to be a useful source of military plutonium (for use in weapons) as well as electricity. Two aspects of its design allow this. Like British nuclear reactors, it has an on-load refuelling capacity, which means that operators can replace fuel in the reactor core while it is still running, by using an overhead crane gantry and localized containment. But, more importantly, it uses low-enriched fuel. Most commercial nuclear power stations use uranium enriched to 3%, but the graphite-moderated RBMK can operate with fuel enriched to 1.8%. The on-load refuelling and low-enrichment, taken together, allow weapons plutonium to be produced at the same time as electricity - nuclear technology's benign and malign faces showing at the same time!

Chernobyl was known to be one of the largest of the 18 nuclear plants in the Soviet Union (at the end of 1985), producing 15 % of the country's total nuclear electric power capacity of 26,840 megawatts. It was vital to Russian power supply.

The accident surprised Russians and westerners alike, the more so because only two months earlier, an eight page article in the February US edition of *Soviet Life* described it as totally safe. "Even if the incredible should happen, the automatic control and safety systems would shut down the reactor in a matter of minutes", it noted. Mr Vitaly Sklydarov, the Ukrainian Minister of Power, was quoted as saying " the odds of a meltdown are one in 10,000 years". It seemed that now was the time to recalculate the estimates of risk at the "ultra-safe " plant!

NUCLEAR SAFETY RECORD

Suddenly the eyes of the world were turned towards the accident at Chernobyl, which had clearly shaken the Soviet authorities. Partly to give themselves diplomatic breathing space and partly to play down the event to minimize public alarm at home, they quickly seized on the opportunity to launch a propaganda offensive against what they saw as the nuclear accident prone west. A second *Tass* report, published later on the evening of Monday 28 April, noted that the Chernobyl accident was the first of its kind in the Soviet Union. It added, tersely, that similar accidents had happened on several occasions in other countries in the west.

The nuclear industry world-wide has *not* had a clean bill of health. Its track record is flawed by a series of accidents since Russia became the first country to use nuclear power to generate electricity for commercial purposes (at the Obninsk research centre near Moscow) on 26 June 1954. Some accidents are better reported than others because secrecy is widespread and the amount and type of information made available on domestic nuclear industries is strictly controlled by many Governments (for both strategic and commercial reasons).

Several episodes stand out in the nuclear industry's track record.

Windscale (1957)

What is widely regarded as the worst nuclear accident in the west occurred in 1957 at the Windscale complex (now called Sellafield, see Figure 1.1) in north-west England. Little information on what happened was made available at the time - largely to avoid public panic but also to play down the military implications. It is believed that a graphite fire destroyed the core of a reactor that was being used to make weapons plutonium, releasing a dangerous cloud of radioactive materials which was blown (and fell to ground) over an area around 500 square km in size.

It is estimated that the 1957 Windscale accident leaked between 3,700,000 and 7,400,000 million becquerels (10,000 to 20,000 curies) of iodine-131 and around 370,000 million becquerels (1,000 curies) of caesium-137. Medical experts are divided over the likely effects of the accident on human health, but some have argued that it could have caused up to 33 extra deaths from cancer over the following decades. Certainly there is evidence of an increased incidence of leukaemia amongst local children, even today.

Kyshtym (1957-8)

Western nuclear experts were not persuaded by the Monday evening *Tass* statement, and insisted that there had been up to three nuclear accidents in Russia since 1954. By far the worst was the huge explosion which occurred near the town of Kyshtym, just east of the Ural Mountains, during the 1957-8 winter. This particular accident is shrouded in secrecy, and details only emerged 18 years after the event via dissident scientist Dr Zhores Medvedev. It is believed that the CIA monitored it, notified the Atomic Energy Authority in Britain and told them *not* to make the news public in case it stirred up public opposition to nuclear power, then in its infancy with - it was hoped - a golden future ahead.

Details of the Kyshtym accident are patchy. It was associated with buried nuclear waste. Medvedev describes what happened: " suddenly there was an enormous explosion, like a violent volcano. The nuclear reactions had led to overheating in the underground burial grounds. The explosion poured radioactive dust and materials high up into the sky. " An area up to 1,000 square km in size was badly contaminated with radio-active fallout, and there must have been some loss of life from radiation sickness (some believe hundreds might have died). At least 30 villages were abandoned and subsequently deleted from Russian maps.

Three Mile Island (1979)

The worst US nuclear accident was at the Three Mile Island plant near Harrisburg, Pennsylvania, in March 1979. The most serious problem in a nuclear reactor is the risk of a core melt-down, which would be uncontrollable and cause massive environmental damage (although the so-called '*China syndrome*' - in which the molten reactor melts its way into the earth's crust - is now widely regarded as a science fiction myth rather than a real scientific possibility).

The partial melt-down at Three Mile Island was the closest 'near-miss' yet. Local residents were evacuated whilst the reactor was sealed and the problem brought under control. Little radiation escaped from the plant because most was trapped within its containment building; it was apparently impossible to detect radiation more than a few kilometres from the plant. The best estimates of radiation leakage put it at around 11 million becquerels (3 curies) in total.

These three are the worst but by no means the only nuclear accidents to occur before Chernobyl. Other serious ones include the 1975 fire in the control room of a nuclear reactor near Athens in Alabama (USA); a melting of some fuel elements in a reactor at Santa Susanna in California (1959); an explosion in a reactor at Idaho Falls (USA) in 1961; partial core melt-down in the Enrico Fermi reactor in Detroit (1966); the more recent (1983) accidental release of radioactive waste into the Irish Sea from the Sellafield plant; and near melt-down at the French reactor at Bugey in April 1984. Between 1971 and 1985 there were over 150 nuclear accidents of varying degree of seriousness in 14 different countries.

Some argue that this track record is not at all bad, given the large number of reactors in use around the world today. It is estimated, for example, that by early 1986 the world's operating nuclear plants had clocked up over 3,800 reactor years of operation (over 1,000 in the USA, 700 in the United Kingdom and over 550 in the USSR).

Chernobyl in perspective

Early estimates of the quantity of radioactivity released from Chernobyl put this accident way ahead of the rest, and quickly earned it the dubious title of *"worst accident in the history of nuclear power"*. The cloud of material blowing out of Chernobyl was by now known to contain non-volatile fission products, indicating that the reactor fuel core must have broken up in the intense heat of the fire. This made the accident decidedly worse than any previous ones.

A report in *Nature* on 8 May suggested that Chernobyl may have released up to a hundred times as much radiation as the 1957 Windscale fire released. This figure was in the course of time to rise dramatically. Chernobyl is now estimated to have released around 18,500 million million becquerels (50 million curies) of radiation - something like 2,500 times as much as the 1957 Windscale fire, and 16 million times as much as Three Mile Island. The US nuclear safety expert who commented on 29 April that Chernobyl made Three Mile Island "look like a tea party" was making the understatement of the day!

THE DAY THAT SHOOK THE WORLD

One thing was certain on that Monday evening (28 April 1986) - the accident at Chernobyl was big; it was serious and it was quite possibly still pouring out radiation uncontrollably. It had been a hectic day of monitoring, questioning, reflecting on what it all meant. Countries throughout Europe were keeping a watchful eye on results from their own monitoring stations. Norway, Sweden, Finland, Denmark and West Germany - all of whom had detected high levels of radiation over the day - were particularly anxious.

It was a source of great political concern that the first signs that anything untoward had happened in Russia were actually picked up in Sweden. The initial Soviet reluctance to acknowledge that a nuclear accident had occurred, and its concession only after strong diplomatic pressure and world-wide news coverage, won Russia few friends in the west. It was particularly ironic that Sweden detected the nuclear fallout from Chernobyl first, because Sweden is decidedly anti-nuclear. Although nearly half of the country's electricity was being produced from twelve nuclear power stations by 1985, a 1980 referendum had voted to phase out the nuclear plants by the year 2010.

The nuclear genie had been well and truly let out of the bottle at Chernobyl. But the questions occupying the minds of people in the west were "*Where will the radioactive fallout land*?" and "*What risk is there to humans*?" The first was in the gift of the weather system - winds would certainly blow it over a large area, and rains would certainly wash it out of the sky onto fields, lakes and rivers, and - the biggest immediate threat - onto towns and people. But where, when and in what quantities? The world could only wait and hope (see Chapter Four).

The risk to humans is a real issue, and fear of radiation poses a very serious psychological problem. Radiation is probably the world's biggest invisible menace; we cannot hear it, feel it, touch it, taste it or smell it. Many people see the prospect of Nemesis (the Greek goddess of retribution) coming alive through nuclear accidents which leak radiation over a wide area and over innocent people. But the fear of radiation is well founded, because radiation is known to be associated with various forms of cancer in humans. We need to examine some of the evidence about that, in the next chapter, before we can start to fully appreciate the significance of Chernobyl.

2
RADIATION AND HEALTH

As soon as the news broke that there had been an accident at a nuclear power station, with radiation leaking out uncontrollably and being spread by the wind over a massive area, nuclear health experts in the west started to gather their thoughts on how it was likely to affect humans both inside and beyond Russia. In the absence of firm advice or information from the Russians, they could only give informed guesses based on assumptions about the type and amount of radioisotopes which might have leaked out from the damaged reactor.

The signs were not good. Early measurements over Sweden and elsewhere suggested that it was likely to be a rather sinister cocktail of fission products - some very long lasting (both in the environment and in human bodies), and many known to accumulate in body tissues (like caesium-137) or to cause bone cancer (like strontium-90).

The early evidence suggested that many people close to Chernobyl were likely to die quickly of radiation poisoning, and a great many more were likely to die in the months and years ahead from cancers induced by the high radiation levels. Whether radiation levels over Europe were high enough to cause additional cancer deaths outside the Soviet Union was at this stage totally unknown.

But this was largely uncharted territory, because the experts have little direct experience of the epidemiological links between people and radiation (especially given the limited number of serious accidents in power stations previously (see Chapter One)). Assessment of likely health risk had to draw on available evidence. This comes from a variety of sources, including the survivors of the atomic bombings at Hiroshima and Nagasaki in 1945, and Marshall Islanders (in the Pacific) exposed to radiation from US military nuclear tests during 1954; hospital patients undergoing X-ray treatment; doctors and staff in X-ray departments, uranium miners and nuclear reactor workers (with high occupational exposures); and people living in areas with high natural radioactivity.

Paradoxically, the nuclear health experts would be wiser and better informed by monitoring the human health consequences of Chernobyl. But the initial speculations had to be based on what was known about background levels of radiation, and about how different levels and types of radiation affect human health.

BACKGROUND RADIATION - SOURCES AND LEVELS

Many people believe that all radiation in the environment is man-made, and that most of it comes from military testing of nuclear weapons and from leaks from civil nuclear installations. The evidence tells us otherwise - we receive radiation from both artificial and natural sources.

Radiation budgets

The amount and composition of these different sources of radiation vary from place to place, but generalized estimates are available. International Atomic Energy Agency figures suggest that just over two thirds (67.6%) is natural background radiation, and we receive most of the rest (30.7%) from medical sources (like X-rays). Less than a sixtieth (a mere 1.7%) comes from all other sources, including the nuclear industry and military testing. Detailed studies by the National Radiological Protection Board show that in Britain some 87% is background radiation, and a further 11.5% is medical (Figure 2.1). Fallout from weapons testing accounts for a mere 0.5%. The nuclear industry contributes a thousandth (0.1%) of the background level in a typical year in Britain.

Attempts have been made to estimate budgets for radiation exposure of people in Britain and the United States (Table 2.1). The estimates look extremely accurate and precise, but they are general figures which mask large variations from place to place. At least the relative magnitudes shown are believed to be realistic. A widely accepted figure for the natural background radiation level is around 2 millisievert a year, equivalent to about 0.5 microsievert an hour (see Table 1.1 for units), at normal altitude and in mid-latitudes (like Britain).

Natural sources of radiation

Most of the radiation to which we are exposed on a day-to-day basis comes from entirely natural sources - the air we breathe, the sun and outer

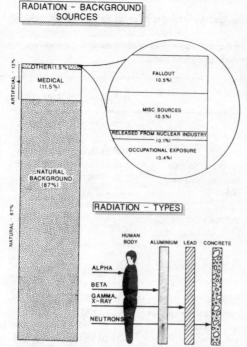

RADIATION – BACKGROUND SOURCES

RADIATION – TYPES

Figure 2.1   Radiation - background sources and types.
Sources after Hughes & Roberts (1984), types after Wright &
Prentice (1986)

space, and rocks and soils (even building materials). These give us
around 1.87 millisieverts of radiation a year (87% of the total received)
in the UK and 1.08 millisieverts a year (57%) in the USA (Table 2.1).

Much of the natural background radiation (probably around 15% of
the total received in both the UK and the USA) comes from the sun and
from outer space, in the form of cosmic rays. The size of this so-called
'*extraterrestrial component*' varies across the earth's surface. It increases with latitude, and exposure at the equator is about 10% lower
than in Britain. It also increases with altitude, doubling with each 1,500
m rise above sea level (in the lower atmosphere). So people living in
Boulder, Colorado (over 1,500 m above sea level) may receive up to twice
as much cosmic radiation as people living on the low-lying east coast of
the USA.

Natural sources on earth also contribute to the background radiation.
This '*terrestrial component*' (around 19% of the total received in the UK
and 13.5% in the USA) comes mostly from radioactive substances in the
earth's crust (over 90% on average comes from   thorium-232   and

19

Table 2.1 Background sources of radiation in the UK and USA

Summary of the average annual dose to members of the population, expressed in millisievert per year (rounded figures).

source		UK (1984)	USA (1980)
NATURAL		1.87 (87%)	1.08 (57%)
	cosmic	0.30 (14%)	0.28 (15%)
	terrestrial gamma	0.40 (19%)	0.26 (13.5%)
	internal irradiation	0.37 (17%)	
	radon	0.70 (32%)	
	thoron	0.10 (5%)	
	internal - gonads		0.28 (15%)
	- lungs		0.26 (13.5%)
ARTIFICIAL		0.28 (13%)	0.83 (43%)
	medical	0.25 (11.5%)	0.78 (40.4%)
	occupational exposure	0.01 (0.5%)	
	miscellaneous	0.01 (0.5%)	
	fallout	0.01 (0.5%)	0.05 (2.6%)
	nuclear industry	0.002 (0.1%)	< 0.01 (0.5%)
TOTAL (rounded)		2.15 (100%)	1.91 (100%)

SOURCE: UK data from Hughes & Roberts (1984), US data from National Research Council Committee on Biological Effects of Ionizing Radiation (1980)

potassium-40 in rocks and soils). The level of radiation from this source varies from place to place. Some rocks and soils emit more radiation - largely in the form of radon gas - than others. Granite areas produce particularly high levels of radon, so that those who live in granite areas may be exposed to several times more radiation than they would elsewhere. People living in Aberdeen in eastern Scotland receive an estimated 1 millisievert a year from natural terrestrial sources alone. No detectable health problems have yet been associated with such small-scale natural variations in radiation exposure.

A third source of natural background radiation is *internal*, because it comes from the material within our own bodies. Some isotopes are formed naturally in human tissues; it is estimated that this source might account for up to 17% of the average total dose of people in Britain (Table 2.1).

Artificial sources of radiation

The artificial sources provide in the region of 0.28 millisievert a year (13% of the total received) in the UK and 0.83 millisievert a year (43% of the total) in the USA (Table 2.1). Almost all of this comes from medical sources (including diagnostic X-rays and radiotherapy). The rest is accounted for mainly by very low levels of occupational exposure (in industries which use ionizing radiation), and miscellaneous sources including air travel, television and luminous watches. Fallout from weapons testing and leakage from the nuclear industry contribute almost nothing to the radiation levels to which most people are normally exposed.

These figures can be set alongside estimates of the amount of radiation we are exposed to in other ways. For example, one chest X-ray may give around 0.20 millisievert, or nearly 10% of the average total annual exposure of a person in the UK. One hour of flying in a jet plane would give around 0.0085 millisievert (8.5 microsievert), so that a 6 hour trans-Atlantic flight in a jumbo jet would give more radiation than the average person receives from nuclear fallout in the USA in an average year. One hour of flying supersonic would give about 0.016 millisievert (16 microsievert), so a 3 hour Concorde flight from London to New York would give nearly as much as the 6 hour flight by jumbo.

Recommended safety limits

The International Commission on Radiological Protection (ICRP) has advised that ordinary members of the public should not be exposed to unnecessarily high levels of radiation from artificial sources, to avoid health risks. Its recommended upper limits are 5 millisieverts in a given year, and a life-time average of 1 millisievert a year. People in Aberdeen need not panic, because the figures exclude radiation from natural (eg rock) and medical sources. But this does highlight the wide range of variations in natural background levels of radiation to which we are exposed, and which place some of us alarmingly close to the recommended safety limits.

Radiation workers form a separate group who are exposed to high levels of occupational exposure to radiation and for whom the recommended upper limits are 50 millisieverts a year. This is because their health can be closely monitored, through regular medical checks and radiation counts. It is assumed that anyone who develops early symptoms of radiation sickness within this high risk group will be detected early enough to allow them to switch jobs (to avoid any further exposure which could be damaging to their health).

The relatively low levels of radiation to which most of us are normally exposed pose little known danger to human health. Certainly when viewed alongside other factors which might cause cancer and genetic damage, the nuclear industry - under normal conditions - poses little real threat. The industry claims, with some pride, that remarkably few people have died as a direct result of radiation from civil nuclear installations, especially given the large number of reactors now in use around the world. Yet, when something does go wrong, and radiation is released into the environment, the risks rise immediately.

RADIATION DAMAGE TO HEALTH

Medical experts are agreed that damage to health from exposure to radiation depends on three things - the type of radiation emitted, the dose of radiation which the body receives, and the specific organ or tissues in which the radioactive material concentrates.

Radiation type

Atoms of certain elements (the so-called *'radionuclides'*) spontaneously emit minute radioactive particles and/or electromagnetic radiation. The particles may have an electrical charge (positive or negative), or they may be neutral (ie have no charge). Electromagnetic radiation is neutral. There are four types of radiation, and each is capable of penetrating different materials in different ways (see Figure 2.1):

Alpha radiation consists of positively charged particles emitted by the atoms of elements like uranium and radium. It cannot pass through dense materials; it can just penetrate the surface of skin, and a sheet of paper will shield it out altogether.

Beta radiation consists of negatively charged electrons which are much lighter than the relatively heavy alpha particles. These can penetrate more and further than alpha particles; they can pass through paper and skin, but are stopped by a few millimetres of aluminium.

Neutrons are uncharged heavy particles which exist within the

Table 2.2 Physical characteristics of important radioactive isotopes

name		radiation	
		type	max energy (MeV)
carbon-14	*	beta	0.16
caesium-137	x	beta	0.18
		gamma	0.66
iodine-129	x	beta	0.15
		gamma	0.04
iodine-131	x	beta	0.81
		gamma	0.72
krypton-85	x	beta	0.67
		gamma	0.52
phosphorus-32		beta	1.7
plutonium-239	x	alpha	5.1
potassium-40	*	beta	1.3
		gamma	1.5
radium-226	*	alpha	4.8
		gamma	0.64
radon-222		alpha	5.0
		gamma	0.5
strontium-90		beta	0.54
tellurium-132		beta	0.22
		gamma	0.22
thorium-232	*	alpha	4.0
		gamma	0.06
tritium beta (=hydrogen-3)	*	beta	0.02
uranium-233		alpha	4.8
		gamma	0.1
uranium-235		alpha	4.6
		gamma	0.39
uranium-238	*	alpha	4.2
		gamma	0.05

* primary isotopes that contribute to background radiation
x isotopes produced in nuclear fission reactors

SOURCE: summarized from Ehrlich *et al* (1977)

nucleus of all atoms heavier than hydrogen. They are very penetrating, and can only be shielded out by concrete or substances which contain hydrogen (like water or wax).

Gamma rays are a form of electromagnetic radiation (similar to X-

Table 2.3   Half-life of radioactive isotopes

Radioisotopes tend through time to revert to their natural form by giving off radiation. This process is known as '*radioactive decay*', and each isotope has its own rate of decay, expressed as its '*half-life*' (the time taken for the radioactivity to fall by half). Isotopes with long half-lives remain in the environment (and pose a threat to humans) for a long time; those with short half-lives disappear fast.

half-life	isotope	
30 seconds	rhodium-106	
17 minutes	praseodymium-144	
57 minutes	rhodium-103	
72 minutes	tellurium-129	
40 hours	lanthanum-140	
3.3 days	tellurium-132	
5.3 days	xenon-133	
8.1 days	iodine-131	x
12.8 days	barium-140	
13.8 days	praseodymium-143	
32.5 days	caesium-141	
35 days	niobium-95	
40 days	ruthenium-103	
52 days	strontium-89	
65 days	zirconium-95	
1 year	ruthenium-106	
1.6 years	cerium-144	
2.1 years	caesium-134	
2.26 years	promethium-147	
2.4 years	neptunium-239	
9.4 years	krypton-85	x
12.3 years	tritium (hydrogen-3)	*x
29 years	strontium-90	x
30 years	caesium-137	
89 years	plutonium-238	
1,622 years	radium-226	x
5,730 years	carbon-14	*
24,400 years	plutonium-239	x
210,000 years	technetium-99	
900,000 years	zirconium-93	
2,100,000 years	neptunium-237	
17,000,000 years	iodine-129	x
1,300,000,000 years	potassium-40	*
4,500,000,000 years	uranium-238	*
14,000,000,000 years	thorium-232	*

> *   primary nuclides that contribute to background radiation
> *   isotopes produced in nuclear fission reactors

SOURCE: summarized from Priest (1973)

rays, light waves and radio waves), which are emitted by the nucleus. They can penetrate much further than the alpha and beta particles, and they require lead slabs, thick concrete or water to shield them out.

The amount of damage caused by radiation is expressed in terms of the sievert (formerly the rem; see Table 1.1). It depends mainly on the radiation dose received, as expressed in grays (previously in rads). The sievert and the gray are related but not interchangeable, because some types of radiation are more effective than others in producing biological damage. Thus 1 gray of one form of radiation might produce much more biological damage than the same amount of another.

The RBE (short for *Relative Biological Efficiency*, sometimes called the *Quality Factor*, or QF) is a measure of the potential of a given type of radiation to cause biological damage. It is expressed relative to the damage potential of an X-ray. X-rays, gamma rays and beta rays have an RBE of 1, natural alpha particles have an RBE of 10 (ie they cause ten times the damage), and heavy nuclei and fission particles have an RBE of 20 (they are twice as damaging as the alpha particles). The sievert is equal to the gray multiplied by the RBE, so that it provides a measure of potential harm to humans which integrates level and type of radiation involved.

Alpha particles pose a much greater threat to health than the beta and gamma rays, measure for measure. Different radioisotopes emit different types and levels of radiation (Table 2.2), so they don't all pose the same risk to health. Those which emit large amounts of alpha particles - like plutonium-239, radium-226, radon-222, thorium-232 and the uranium isotopes - are a much greater problem than the rest.

Radiation dose

Health problems also depend on the dose of radiation which the body receives, and this reflects three things. One is the amount and type of radiation released at source. But since the amount of radiation emitted by radionuclides decays through time, the dose received - certainly for those elements with a relatively short half-life (Table 2.3) - will also depend on the timing of exposure relative to release.

The third control is the way in which a person comes into contact with the radiation. People close to the Chernobyl reactor when it exploded will have been exposed directly to high levels of radioactivity. Whole body external exposure in those cases would quickly lead to severe radiation sickness and quite possibly death. But there are a number of other pathways along which radiation reaches humans - including external

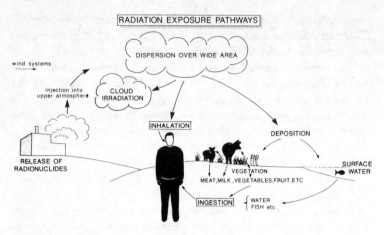

Figure 2.2   Radiation exposure pathways to humans

exposure to material in the sky and clouds, and internal exposure via breathing (inhalation) and eating and drinking (ingestion) (Figure 2.2). Each of these exposes the body to different levels and types of radiation, depending mainly on where we live and work, and what we eat and drink.

If there is a serious threat of radiation contamination over a wide area (as, indeed, there clearly was from Chernobyl), individuals a long way downwind of the source can be exposed to quite high levels of radiation through the simple act of breathing. Staying indoors, away from contaminated air, can help to reduce risk. Similarly, individuals can be exposed by drinking contaminated milk and water, and eating contaminated food (meat, fish and fresh vegetables). Careful control over what is eaten and drunk can minimize exposure risk and thus protect health. This control can be voluntary (for instance, an individual might switch to tinned food and dried milk for a period) or it might be imposed and enforced by the authorities (for example by banning the outdoor grazing of sheep and cattle whilst ground contamination is high).

The International Atomic Energy Agency has set 'emergency levels' for concentrations of iodine-131 in milk. These are 1,000 becquerels a litre for children, and 10,000 becquerels a litre for adults. This does not necessarily mean that milk with higher levels will directly endanger health - the emergency levels are used to ban the drinking of milk as a preventive measure, to reduce the risk of long-term cancers.

Damage and radiation dose are closely linked. The higher the radiation levels, the greater the damage to health. But there are a series of critical radiation levels at which different symptoms appear (Table 2.4).

Table 2.4 *Medical effects of radiation in humans*

dose (gray)	symptoms	deaths (average)
0-1	no long-term observable effects, men become temporarily sterile in the range 0.2-0.5 gray.	0
1-2	nausea and vomiting within 3-6 hours of receiving dose and lasting less than 1 day, followed by no symptoms for 2 weeks. Recurrence of symptoms for another 4 weeks. Number of white blood cells reduced.	0
2-6	nausea and vomiting lasting 1-2 days. No symptoms for 1-4 weeks, followed by recurrence of symptoms for up to 8 weeks. Diarrhoea, severe reduction of white blood cells, blood blisters on skin, bleeding, infection. Loss of hair above 3 gray.	0-98% in 2-12 weeks from internal bleeding or infection
6-10	nausea and vomiting starting within half hour of receiving dose of radiation and lasting 2 days. No symptoms for 5-10 days then same symptoms as for 2-6 gray for 1-4 weeks.	98-100% from internal bleeding or infection
10-50	nausea and vomiting starting within half hour of receiving dose and lasting less than a day. No symptoms for about 7 days, then diarrhoea, fever, disturbed salt balance in blood for 2-14 days.	100% within 14 days from collapse of circulation
over 50	nausea and vomiting immediately followed by convulsions, loss of control of movement and lethargy.	100% within 14 hours from failure of breathing or brain damage

NOTE: effects apply for a normal population of adults receiving total radiation doses over short periods of time, and may vary very widely in the 1-6 gray range

SOURCE: Greene (1982)

These make it possible to predict how the health of an individual person might be affected who has been exposed to a known level of radiation. They also allow estimates to be made of the likely number of deaths in a population which has been exposed to a known level of radiation (like the

Chernobyl

residents of Pripyat and Chernobyl, close to the damaged nuclear reactor).

The average person faces no real health problems from exposure to normal levels of background radiation. Health effects only start when radiation levels rise above about 0.05 sieverts, and at this low level some cancers appear but no one dies directly (Figure 2.3). Once the radiation dose exceeds about 1 sievert early mortalities (for example through bone marrow failure) can be directly related to radiation poisoning.

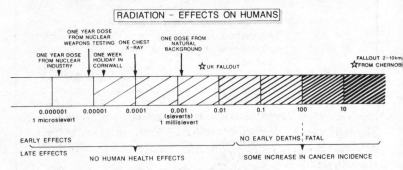

Figure 2.3 Radiation - effects on humans according to dose.
Based on information in Serrill (1986) and Wright & Prentice (1986)

While it is possible to talk of increasing incidence of cancer and related medical problems, it is impossible to identify the individuals who are most likely to be affected within the overall population which has been exposed. This is yet another dimension to the sinister game of Russian Roulette!

Organ or tissue involved

Health damage also depends on the specific organ or tissues in the human body in which the radioactive material concentrates. This depends almost entirely on the character of the radiation involved; the level of radiation and the way in which we are exposed are of less significance.

We noted above that alpha particles have a high RBE, but few of the radioisotopes likely to be released from a damaged fission reactor emit alpha radiation (Table 2.2). However, different isotopes affect different organs (Table 2.5) - some attack the whole body, while others accumulate in the thyroid or lungs. Health problems thus reflect both the cocktail of isotopes to which humans are exposed (which determines the organs and tissues to be affected), as well as the overall level of radiation

Table 2.5 Radioisotopes and the organs they affect

isotope	organ affected
caesium-137	entire body
iodine-129	thyroid
iodine-131	thyroid
krypton-85	lungs, skin
plutonium-239	entire body, especially lungs
tritium (hydrogen-3)	entire body
NOTE: all of these isotopes are produced in nuclear fission reactors	

SOURCE: Tyler Miller (1979)

involved (which affects the amount of damage to that organ or tissue). The early fears surrounding Chernobyl fallout centred on high levels of caesium-137 and of iodine-131; both are serious threats to human health, and both can kill in different ways.

TIMESCALES OF RISK

The grim prospect of large numbers of casualties began to turn over in the minds of the nuclear health experts as they listened to the early reports about Chernobyl. They more than any others were aware that it will never be possible to count the total human cost of the accident, because many of the medical problems arise years if not generations after the event. Radiation affects human on three timescales:

Acute effects will only affect people who were close to the source of radiation (in this case the power station when the explosion occurred). This is the most critical group. Many - some experts anticipate up to a half - are likely to die within the first five to ten weeks, and the rest are likely to develop cancer of one form or another within a matter of years.

The immediate effect of exposure to very high levels of radiation is destruction of the bone marrow, through irradiation. This in turn destroys the body's immune system, leaving the body vulnerable to infection (much like AIDS does). The first deaths (within the first four weeks) are likely to be from infection (such as through pneumonia), and others will die in the next few weeks from liver or kidney failure. People with bone marrow failure are almost certain to die within days or weeks unless bone marrow transplants are carried out (see Chapter Four) and successful.

Long-term effects will affect people living further away from the source, who will none the less have been exposed to relatively very high radiation doses. They will also take their toll on the people who were close to the source and managed to escape the worst ravages of the acute effects. Fewer are likely to die directly from cancer, but cancer and chronic illness will still be widespread. Many will develop cancer decades after they receive the high doses - this is sometimes called the *Hiroshima Syndrome* - so that rates of incidence of leukaemia and cancers of the thyroid, lungs and breast are likely to rise 15 to 20 years after exposure (ie between AD 2001 and 2006).

The fertility of many affected people will also be impaired, so they will not be able to have any (more) children. Those who do not become infertile could quite possibly give birth to children suffering from genetic and hereditary diseases and/or severe mental handicap or retardation. Mothers in the second to fourth month of pregnancy at the time of exposure are the highest risk group.

Genetic effects may occur in the children and subsequent generations of those in the 'long-term effects' group who were not made infertile. Their heirs might be affected in a variety of ways, including death in early childhood and emergence of many hereditary conditions in adults and the elderly. But these are almost impossible to predict because they may not appear for many generations to come. The long-term price may indeed be very high.

FEARS AND FACTS

There is a bewildering array of possible health consequences of a major nuclear disaster like what appeared to have happened at Chernobyl. As the UK House of Commons Environment Committee concluded in its 1976 report on radiation, "for most people radiation is inexplicable, unseeable, untouchable and almost mystically evil". The unfamiliar units of measurement used, the difficulty of knowing what it all implies, and the widely discussed links between radiation and cancer (the two major evils of our time) only serve to further heighten public anxiety.

Swedish measurements suggested that radiation levels in the skies over Scandinavia were highest on 29 April, up by between 10 and 20 times the norm. Most of the radioactivity appeared to be coming from iodine-131 and caesium-137, and measures were swiftly introduced to minimize the health risks these pose to humans (initially in the Nordic countries, but progressively across Europe in the days that followed (see Chapter Four)).

People in the west hoped that they would be fully informed what the risks were - by the Russians and by their own governments. But this would take some time, because of the lack of information in and from the Soviet Union on the composition and size of the radiation leak. Indeed, it was to be some time (12 days) before the damaged reactor was sealed and radiation stopped pouring out directly into the atmosphere. Before that could happen, the Russians were fully engaged in a desperate struggle for containment at Chernobyl.

3
STRUGGLE FOR CONTAINMENT

Eyes around the world were focussed on the Chernobyl area and on Soviet attempts to cope with the disaster. Foreign diplomatic pressure, coupled with an intent to play down the seriousness of the accident at home, forced the Soviet government to issue a further statement through *Tass* on the Tuesday evening (29 April).

Like the first one it was brief and - to many western observers - somewhat unilluminating. It spoke of two dead and an undisclosed number of people evacuated from the area; it added that the " radiation situation " at the plant was now stabilized and "necessary medical aid " had been given to those affected by the explosion. "Priority measures have been taken to deal with the effects of the accident. The inhabitants of the station's settlement (Pripyat) and three nearby populated localities have been evacuated. "

Few details were given of the accident, which " resulted in the destruction of part of the structural elements of the building housing the reactor, its damage, and a certain leak of radioactive substances". Only the fourth reactor had been affected, but the other three had been shut down as a precaution. People in Russia and beyond were still very much in the dark about what was happening at Chernobyl, and what the dangers were both there and downwind. Was the graphite fire still raging out of control? What was the prospect of a melt-down?

OFFERS OF HELP

Diplomatic differences and propaganda potential were set aside, if only briefly, by many western countries who recognized the urgency of the situation and made unconditional offers of help. The United States and West Germany offered technical and medical experts and equipment;

Japan and France offered to treat those already suffering from radiation poisoning.

The Soviet authorities' main concern was to control the fire. In an unprecedented move (believed to be the first time since 1917 that Russia has asked another country for help in time of disaster) they quickly turned (on Tuesday 28 April) to nuclear experts in Sweden and West Germany for guidance. After all, up to that point most discussions anywhere in the world about how to cope with serious radiation leaks - civil or military - had been essentially theoretical; practical experience was mercifully thin on the ground.

The Swedish Nuclear Safety Inspectorate advised them to ask Britain - through the UK Atomic Energy Authority (UKAEA) - for help, because of its experience of successfully fighting the graphite fire at Windscale in 1957. Britain had anticipated the prospect of being able to help, and immediately put a crisis team at the disposal of the Soviet authorities. The team comprised experts from the UKAEA, British Nuclear Fuels Ltd (BNFL, which operate the Sellafield reprocessing plant), the National Radiological Protection Board (NRPB) and the Nuclear Installations Inspectorate (NII).

However, as some journalists soon pointed out, there were few parallels between Chernobyl and Windscale. The 1957 fire could be extinguished by flooding the containment building with water; there was no containment building as such at Chernobyl, and pouring water onto the burning graphite reactor there would only create a cloud of radioactive steam and make the fire more intense (because of the large so-called '*positive void coefficient*' of the Chernobyl reactor design - see Chapter Seven).

Moscow rejected all unsolicited offers of help from other countries, and - as if to add insult to existing injury of the international community - asked none for assistance. Later that week, however, mounting concern over the prospect of many deaths from bone marrow failure amongst those exposed to the highest levels of radiation in the Chernobyl power station made it impossible for the Russians to refuse medical help from US bone marrow specialist Dr Robert Gale (see Chapter Four).

The Soviet Union also accepted gifts of equipment from abroad. West Germany sent two remote-controlled robots which made it possible to film inside the damaged reactor, which was still highly radioactive. These proved invaluable to the Russian scientists who were trying to assess the extent of damage to the reactor, and decide how best to deal with it. Britain also sent equipment. Over £150,000 worth of fire fighting and personnel decontamination equipment went from BNFL (the operators of Sellafield/Windscale), and the Central Electricity Generating Board (CEGB) sent 50 protective hot suits.

THE ACCIDENT - POSSIBLE CAUSES AND CONSEQUENCES

The lack of definitive information forthcoming from the Russians did little to dampen the enthusiasm with which western nuclear specialists offered their guesses as to what might have caused the accident at Chernobyl. There seemed to be two ways of accounting for the accident. It could be the result of a technical fault - such as a failure of the station's control systems (which are designed to eliminate fluctuations in activity levels across the reactor core) to work properly. The alternative was a human error - such as an operator in the control room hitting the wrong button and causing the control rods to withdraw from the reactor core prematurely. Most hedged their bets and opted for a combination of human error and mechanical failure. At this stage no-one, not even the Russians, knew for sure.

Preliminary analyses of the fallout over Sweden, Finland and Denmark had established that at least some of the fuel from the reactor core must have melted, because it contained substances like polonium which are not very volatile. Such material could only come from a hot vapour given off by a molten pool of fuel. There must have been a serious fire involving graphite and fuel materials in at least one of the station's reactors. That much was fact; the rest was at this time informed speculation.

The likely sequence

Clearly, all authorities agreed, fuel in the reactor must have overheated and then caught fire. The most likely cause would be failure of the cooling system in the reactor core, which would trigger off an almost instantaneous series of reactions (Figure 3.1). Loss of coolant would cause activated fuel in the core to get hotter and hotter. The metal cladding of the fuel pins would probably melt, rapidly spreading heat throughout the reactor core and causing yet more fuel to burn. The pressure tubes of the failed cooling system would be ruptured, releasing any remaining cooling water as superheated steam. This would react with the graphite moderator to cause a volatile release of hydrogen and carbon monoxide. Result - a violent chemical explosion which would rupture the shielding around the reactor core and tear open the reactor building. This in turn

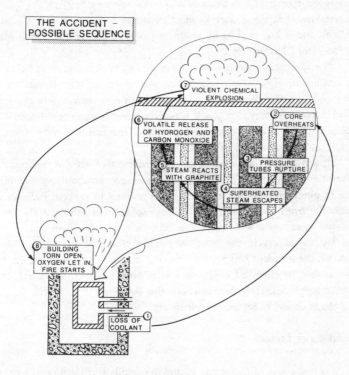

Figure 3.1   The accident - possible sequence

would let oxygen in to the core, and allow the fire to start.

There was over 200 tons of graphite in the reactor, and this was likely to burn for a very long time; indeed, it was likely to be still burning fiercely, out of control. The first priority on site was clearly to put the fire out to prevent the risk of core melt-down. But the explosion which tore open the reactor building also made it possible for radioactive steam and gases from the damaged reactor to escape into the overlying air, from which they would be blown over vast distances by the wind. The need to seal the reactor and prevent continued release of radiation was equally pressing.

Lack of containment

The problem of radiation leakage was seized upon by western nuclear experts, who quickly pointed out that few Russian nuclear power stations are built with containment facilities - double walled protective buildings designed specifically to trap any accidental release of radioactive gases. Containment is mandatory in most countries in the west (see Chapter Nine), and it had proved its worth in the 1979 accident at Three Mile Island (see Chapter One).

The approach to nuclear safety favoured by most countries is based on containment. The Russian approach was based on ultra-reliability - making sure that there are sufficient built-in automatic safety systems, and suitable contingency plans. Russian nuclear reactors are designed to cope with the worst possible case they could imagine (transverse fracture of a 900 mm turbine pressure collector).

It was already self-evident from Chernobyl that without containment facilities, if something *does* go wrong - no matter how unlikely this might be in purely statistical terms - there is nothing to prevent radioactive material from leaking out into the atmosphere and subsequently being blown about by the wind and deposited over vast areas downwind.

By now experts in the west were fairly clear that the reaction which caused the accident had been a chemical one, not a nuclear one (in the sense that the chemical explosion had released radioactive material into the atmosphere). They were also sure that they were dealing with the risk of '*burn-down*' rather than '*melt-down*'.

Additional factors

The actual cause of the accident, and the details surrounding it, were to emerge many months later, when the Russian Government Commission published its report (see Chapter Seven). An official Soviet statement on 6 May said that the accident was "the result of the coincidence of several highly improbable and therefore unforseen failures". This did little to convince western nuclear specialists, who could find few logical explanations for the fact that back-up independent cooling systems (of which there are many in the RBMK reactor design) were not automatically brought into operation when the primary one (presumably) failed. Control of the reactor could also have been regained by the operators in the plant's control room, who - given the time offered by the graphite moderator and the reactor's built-in monitoring systems - might have been able to manually switch on an emergency water supply. Neither had happened. Moreover, the option to automatically shut down the reactor by dropping the control rods (see Chapter One) had not been followed.

The feeling amongst western specialists was that accidents like this would be highly unlikely under routine circumstances. Two thoughts emerged - it could be related to human error (including wholesale negligence on the part of the operators) during routine operations, or may have been caused during some special operation (like the exchange of fuel elements while the nuclear reactor was still running).

It quickly emerged (in a report in the Soviet newspaper *Sovietskaya Rossiya* on 8 May) that the station was under-staffed during the weekend of the accident; many workers had gone away because of the forthcoming May Day holiday. This might have been a mixed blessing. The good news was that few people were present when the accident happened, so the casualty list (from the explosion and subsequent radiation poisoning) had been minimized. The bad news was key trained personnel were quite possibly absent and crucial decisions had to be made on the spot by an under-staffed and over-stretched supervisory team. This would certainly help to account for the apparent uncertainty over how serious the accident was at the time (hence over whether higher authorities, including Moscow, should be informed straight away).

It was also to emerge later (in a report in *Pravda* on 12 May) that number four reactor at Chernobyl was only operating at about 7% of its full capacity at the time of the accident. Russian RBMK reactors were well known to be very unstable (and difficult to control) at low power, so this might have been a significant factor in causing the disaster (see Chapter Seven).

Other speculations began to emerge within days of the accident. Some centred on the possibility that the power station had been built below standard. These were based on revelations from within Russia that the building programme at Chernobyl had been troubled by serious delays (caused by slow delivery attributed to the country's bureaucratic supply system and overloaded transport networks). It was argued that last-minute efforts to catch target dates (involving substantial over-time working) might have led to corner-cutting and perhaps some oversight of quality control.

Problems on site

While western observers were pondering over the possible cause of the accident, the Soviet authorities were deciding how to handle things on site. Academician Valerie Legasov (Deputy Director of the Soviet Atomic Power Institute) was sent to take charge, and he arrived from Moscow a few hours later. The immediate concern at Chernobyl was to extinguish the reactor fire. Intense heat and high radiation levels made this a difficult and extremely dangerous operation for the people most directly involved.

There was also concern over the prospect of one or more of the three adjacent reactors being ignited and damaged by the fire in number four reactor. Evidence was coming to the United States from both commercial and strategic "spies in the sky' ". Commercial satellite pictures, released around the world on 30 April (five days after the accident), showed two bright red spots beneath a cloud of blue smoke, suggesting that a second reactor was already alight (and likely to be leaking radiation). US intelligence (CIA) satellite photographs, rumoured to be available in the Pentagon the same day, reputedly showed a badly damaged reactor building (with a massive hole in its roof and collapsed walls). Washington was later to discount the suggestion that a second reactor was threatened.

AVAILABILITY OF INFORMATION

Moscow had to clarify the situation, and fast. A report on Chernobyl - clearly intended for both domestic and foreign consumption - was prepared by the Council of Ministers and carried on the main evening television news bulletins in the USSR that same day (Wednesday 30 April). It was illustrated by a photo of the plant, said to have been taken after the accident and showing only the top of number four reactor building destroyed. The report added that radiation levels around the plant were falling, and a team of specialists had been called in to clean up polluted areas. It denied that thousands had been killed, and maintained that two had died and 197 had been treated in hospital (49 of whom had been released after medical examination).

Many western diplomats and scientists were unconvinced by the report and claimed that it had deliberately made light of the problem, to avoid wholesale panic in the Ukraine and elsewhere in the Soviet Union.

News black-out

It was to be a further week before the first Soviet reports on what was happening in and around Chernobyl started to appear. The Kremlin imposed a complete information black-out as soon as news of the accident broke in the west. Kiev - 100 km south of Chernobyl, and the nearest major centre - had been declared 'out of bounds' to westerners on the Monday (28 April). On the Wednesday (30 April) a ban was imposed which prevented all western reporters and diplomats from getting within 160 km of Chernobyl.

The news black-out angered foreign governments, who accused

Moscow of a clumsy attempt to cover up details of the disaster. As the days went by the lack of information was also to give rise to mounting concern within the Soviet Union, amongst its 278 million people. Only two groups of people in Russia seemed to have any information to go on, other than workers at the station, local and state Communist Party officials and the authorities in Moscow, none of whom was saying anything. These were people living in the affected area (including Kiev) and those who could receive it from sources in the west, normally by radio.

There was a widely held belief within Russia, carefully cultivated by state-controlled Soviet reporting of the situation, that western media reporting of the accident was exaggerated propaganda. Some insisted that there had, in fact, been no such accident at all - it was a fabrication being put around by western intelligence (especially the US) to discredit the Soviet Union. Others believed that something must have happened at the nuclear power station, but it was nowhere near as serious as the west was making out.

Critical two day delay

This week of continued uncertainty and lack of hard information did little to heal the serious diplomatic wounds which the accident had opened up. The cause was not that the accident had happened, but that two days had passed before the Soviet Union officially announced it (on April 28) and, only then, after Scandinavia had uncovered the story and blown the whistle on them. Western governments were very critical of this delay. Had they been informed straight away that the radiation leak had occurred, countries downwind could otherwise have been making contingency plans to minimize the health risks to their own citizens (see Chapters Four and Five).

Various reasons were offered for the delay. Mr Vitaly Churkin, from the Soviet Embassy in Washington, stressed to the US Congress (on 2 May) that Moscow had to be sure of the extent of the disaster before it broke the news, so as not to unduly alarm its own people. Dr Georgi Abbaton, director of the US and Canada Institute in Moscow, told Russian television viewers (on 4 May) that Moscow had not immediately informed other countries about the accident because its own experts had insisted that there was no danger of contamination spreading over the Russian border. Mr Alexandr Novogodorov, of the Soviet Organization of Friendship Societies, told reporters (on 8 May) that Moscow had delayed announcing the disaster in order to keep spirits high on May Day;

"it happened on the eve of a holiday ... and we did not want to spoil the celebrations ", he added.

Mr Alexander Lyasho, Prime Minister of the Ukraine, revealed (also on 8 May) that the Politburo in Moscow had been told about the accident within hours on 26 April, but "the measurements at first showed that there was nothing to fear " and the authorities did not appreciate its size or significance until two days later. A similar view had been expressed two days earlier by Mr Boris Scherbina, head of the official government inquiry commission on the accident, who claimed on a television news conference (beamed live to the USA) that local staff failed to appreciate the scale of the disaster when it happened at 1.24 am on 26 April, and that incorrect details had been passed on to the central government in Moscow. "The first information we obtained was not the same (as that) which we obtained when we were in the area - in the area, local experts had not made a correct assessment of the accident ", he said.

Formal reassurances

Amid the growing international diplomatic anger over the delay in announcing that the accident had happened - and only then after Scandinavia had detected the fallout from it and demanded an explanation - Moscow had to appease neighbouring countries. Soviet envoys were given a prepared (and very generalized) statement to pass on to embassies and Governments around the world, ostensibly not for public release. Norway released a partial text which had been delivered by the Soviet Ambassador, Mr Dimitry Polyansky, on Wednesday 30 April. It read:

> "As a result of the accident, a part of building
> constructions in reactor buildings were damaged and a
> certain leakage of radioactive material has taken
> place.
>
> Three remaining energy blocks (ie reactors) have been
> stopped, are undamaged, and represent a usable
> reserve. The radiological situation in areas located
> close to the accident site required partial evacuation
> of the population. Measurements are being taken
> continuously.
>
> Efforts taken allowed stabilization of the
> radiological situation. Additional measures are being

implemented to correct consequences of the accident.

Competent Soviet authorities have registered spreading
of radioactive pollution in western, northern and
southern directions.

Pollution level is somewhat higher than acceptable
standards, but not to a degree which should require
special efforts to protect the population.

In the event we get information which should be of
special interest to your country, we will come back to
this. "

The statement was a mixture of good and bad news. It did offer assurances
over the magnitude of the accident (three of the four reactors were
undamaged, for example), and it conceded that partial evacuation had
been necessary. The attempt to play down the risk of serious radiation
contamination over a wide area was less convincing. Moreover, it left
many questions unanswered. Was radiation still leaking? How much
radioactive material had been released? What was the risk of fires
breaking out in other reactors at the plant?

Further details emerge

The official news black-out was lifted on Sunday 4 May, when Soviet
television showed the first moving pictures of the reactor and surrounding area. Until then an official single still photo showing the damaged
reactor building with a hole in its roof was all that was available. The news
film, shot from a helicopter flying over the area, showed few signs of
activity other than convoys of army lorries bringing in special equipment. Blocks of flats in Pripyat were visibly deserted; it had all the
hallmarks of a ghost town. But the limited prominence given to the story,
which appeared well down the news bulletin, showed an official Soviet
intent to play down the disaster.

Selected details of the disaster were released in a carefully orchestrated manner by the Kremlin, eager to capitalize on the propaganda
offensive. A key part was played by Mr Boris Yeltsin, the new head of the
Moscow Communist Party and a senior Kremlin figure (widely believed
to be a close political ally of Gorbachev), who made himself available for
interviews by the international media (on 5 May) during his visit to the
West German Communist Party Congress being held in Habburg. He

revealed that the radiation leak had almost been stopped and radiation
levels in the area were now falling. "Further leaks from the reactor have
almost been stopped (and) the nuclear cloud is now beginning to
disappear and a new one has not formed ", he stressed.

Yeltsin also reported that the fire at Chernobyl was almost out but still
smouldering. Soviet troops in helicopters had been dropping sand, lead
and boron onto the damaged reactor to absorb neutrons and seal the core
(Figure 3.2), and temperature and radiation levels were being measured
using remote-controlled vehicles.

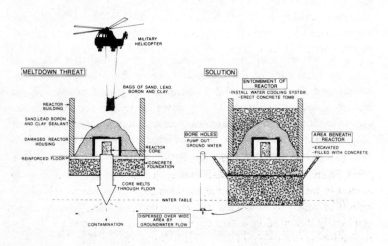

Figure 3.2   The melt-down threat and how it was coped with

Radiation levels on site were now around 100 roentgens an hour (1
sievert an hour, assuming an RBE of 1 (see Chapter Two)), having been
up to 300 roentgens an hour (3 sieverts an hour) on Friday 2 May - both
dangerously high levels which seriously threaten health of anyone
exposed for more than a very limited period. He also announced that
Soviet citizens within 30 km of Chernobyl had been evacuated (see
Chapter Four), the only people still in that area were military personnel
guarding factories, public buildings and property (from looting). No
restrictions had been imposed on selling milk, vegetables or fruit beyond
this tight exclusion zone.

On the same day the Council of Ministers issued a statement in *Tass*
which confirmed much of what Yeltsin had said. It formally acknow-

ledged, for the first time, that radiation had spread beyond the 30 km zone around the plant, but stressed that "the radiation situation on the territory of the Ukraine and Byelorussia is stabilizing with a tendency towards its improvement ". It also revealed that attempts were being made to build up the height of the banks of the Pripyat River beside the power station "to prevent possible contamination ".

Further details emerged the next day (6 May) when *Pravda* carried a report on Chernobyl. It gave details of the accident, in which an initial explosion had destroyed parts of the reactor building and then a fierce fire had started, initially by the engine room. Two explosions occurred one after the other in reactor four at around 1.24 am on the Saturday, throwing out sparks and burning debris. These started around 30 fires around the station, which threatened the neighbouring number three reactor and the machine room adjacent to the twinned reactors. A team of 28 firemen fought the blaze, and had it mostly under control by 5.00 pm that same day.

But it was proving extremely difficult to extinguish the graphite fire in the damaged reactor core, because the high temperatures were causing the water and chemicals being used by the firemen to evaporate immediately.

The Soviet newspaper and television coverage of Chernobyl gave great prominence to individuals whose heroic deeds added human interest to the story and deflected attention away from the more sinister aspects of what was happening. For example, the 6 May *Pravda* report graphically described how "the firemen were fighting the blaze (by the engine room) at a height of 30 m. Their boots stuck in the bitumen that melted because of the high temperature. Soot and smoke made it difficult to breathe. But the brave, bold men kept fighting the blaze courageously ... In the opinion of specialists, the heroic deed of the firemen limited the extent of the accident to a considerable extent ". A Soviet television documentary of the accident and its aftermath, shown on 'Panorama' in Britain on 6 April 1987, also laid great stress on human bravery and dedication in the days and weeks during which many people were involved in containment and clean-up on-site.

THREAT OF MELT-DOWN

The possibility of some form of core melt-down (through the base of the reactor building) could not be overlooked. It was generally agreed that the fission chain-reaction (see Figure 1.3) probably stopped when the explosions occurred. But western nuclear scientists were concerned that the Russian tactic of burying the graphite fire within the reactor beneath a shield of sand and lead - to seal the leak and prevent further release of radioactive materials - would *not* stop the graphite fire ... it might just stop it from cooling down.

Rough calculations suggested that the Chernobyl reactor could still be producing over 10 megawatts of thermal power up to two weeks after the accident, because of heat production from the continued decay of fission products. Whether there was enough energy coming from the entombed reactor to vaporize (quite literally melt) its way through the 1.5 m thick concrete floor to the reactor was a matter of some debate. But the prospect was there.

If that did happen then molten radiation products would come into contact with a series of water-filled pressure-suppression pools located immediately beneath the reactor floor (on top of the 1.5 m thick concrete foundation to the reactor building). This would cause a chain of violent explosions which would cause even more damage to the structure of the power station and perhaps threaten the other one adjacent and two nearby nuclear reactors (which seemed to have escaped serious damage in the first explosions). The prospect was not a happy one!

The only solution was for Soviet engineers to drain the pools - and this involved working unshielded beneath the damaged reactor, where radiation levels were likely to be dangerously high.

Even if this second wave of explosions could be avoided, there was still a very real risk that ground-water would be contaminated (Figure 3.2). This could spread radioactivity into rivers and lakes over a huge area (certainly as far as Kiev, but possibly also throughout the Ukraine and down to the Black Sea), perhaps within the year. Some of the material known to be coming from the damaged reactor had a long half-life (Table 2.3), so it would persist in dangerously high concentrations for decades or longer.

Such a catastrophe would severely threaten the entire future of Kiev, Russia's third largest city with a population over two million. It would also destroy the agricultural potential of most of the Ukraine - an area which traditionally has served as Russia's bread-basket. The social, economic and environmental consequences could be immense and very long-lasting (see Chapter Eight).

It became clear that the area beneath the reactor floor and the building foundation (where the pressure-suppression pools were) must be filled with a protective barrier that would prevent underground contamination. Experiments were carried out at the Atomic Power Institute in Moscow to find what materials would give the best protection. Concrete emerged as the most viable material, so plans were made to fill the area with concrete, up to 5 m thick.

Visit of IAEA experts

Western correspondents were first admitted to the ' *cordon sanitaire*' around Chernobyl when a group were allowed to visit Kiev on 8 May. They were briefed by Mr Alexander Lyasho, the Ukrainian Prime Minister, and accompanied throughout by local officials. This was to be the first independent inspection by westerners since the accident nearly two weeks earlier. They reported that radiation levels in the area were now stable and falling, and that the reactor fire was nearly out.

The journalists had to rely for their information on briefings from informed   Soviet sources and speculations from scientific experts elsewhere. But the opportunity arose for independent external scientific evaluation of the situation when three international experts from the International Atomic Energy Agency (IAEA), a United Nations agency based in Vienna, were invited by Moscow on 4 May to visit the site.

The IAEA team was impartial and international, with impeccable credentials. Dr Hans Blix, Director General of IAEA, was accompanied by Professor Leonid Konstantinou (the Russian Deputy Director General of the Agency's Department of Nuclear Energy and Safety) and Dr Maurice Rosen (American Director of the Division of Nuclear Safety in that Department). They were in Russia from 5 to 9 May and had detailed discussions with Russian nuclear scientists at the headquarters of the State Committee on the Utilization of Atomic Energy.

Highlight of their trip was a brief helicopter flight to within 800 m of the damaged reactor, at a height of 400 m, on Friday 9 May. These few minutes of close observation allowed them to assess the situation for themselves.

The IAEA experts gave the first report of peak radiation levels over the power station site. Their personal dosimeters recorded nearly 3.5 millisieverts an hour - at a height of 400 m, 800 m away from the reactor, six days after the accident. This clearly showed that the area around the plant was still heavily contaminated, and extremely dangerous to work in. A few days of normal shift work in such an environment could easily push workers' body levels into the high risk area (Figure 2.3) (one 8 hour shift would give 28 millisieverts).

They also confirmed and elaborated on what the journalists had been told the day before:

(a) the reactor had been buried under nearly 5,000 tonnes of sand, lead and other materials (which had been dropped onto it from above by Soviet military helicopters) (Figure 3.2)

(b) the fire in the reactor was under control; it was still smouldering but probably out (the colour of the smoke rising from it suggested that sealant material, not graphite, was now burning)

(c) temperatures within the damaged reactor appeared to be falling (but were much still higher than the 300°C claimed by Mr Lyasho)

(d) the initial blast had halted all chain-reactions within the reactor, but it remained hot because the 150 tonnes of radioactive fuel continued to decay; the chain-reaction never re-started, and the risk of melt-down was now minimal

(e) the Soviet emergency team, working under conditions of immense personal risk, was trying to entomb the reactor in concrete from above and below (Figure 3.2).

The IAEA team also reported that the USSR would be keeping the Agency fully informed about developments at the power station, and sending daily radiation measurements from seven monitoring stations around Chernobyl (see Figure 4.1) starting 9 May. IAEA would then send this data on to national bodies concerned with radiation safety in member countries. This was the start of a better exchange of information between the Soviet Union and surrounding countries.

Hans Blix commented after the trip that it was successful in opening up channels of communication with the Soviet authorities, and in allowing external verification of the basis of some of the statements being released in routine announcements in Moscow.

CONTAINMENT

Many of the details given by the IAEA team were further confirmed in *Tass* on 11 May by Mr Vevgeny Velikhov, the Russian scientist in charge of the clean-up. The first official admission that the real threat of melt-down had existed came when he confirmed that it had been averted days earlier, when the graphite fire in the over-heated reactor had been brought under control (12 days after the accident). He also outlined the efforts being made to contain the damaged reactor.

Workers on-site were faced with difficult and dangerous conditions of high radioactivity. Moreover most of the problems they were trying to cope with were new, and they had to be tackled in new and improvized ways. Time was short, risks were high, relevant experience was minimal.

Much of the early effort was directed towards containing the spread of radioactive material, from above and from below. Containment from above was designed to seal the damaged reactor core. Military helicopter pilots dropped sacks of boron, dolomite, lead, sand and clay onto the smouldering reactor, with speed and precision. Each pilot dropped 33 tonnes over 22 missions. Each minute of exposure to such high levels of radiation was critical to their health. By 2 May the damage had been localized around the reactor, and by 6 May the seepage of radioactive materials from the reactor had all but stopped.

Containment from below was designed to help cool the reactor core and so minimize the risk of ground-water contamination. A tunnel 136 m long had to be excavated under the foundations of the reactor b uilding, to allow installation of a giant refrigerator system. Mechanical diggers could not be used, so it had to be dug manually. Volunteer miners from Dombass and Moscow worked in shifts, in difficult and dangerous conditions. They had it finished by the end of May, and then an insulating layer of liquid nitrogen refrigerant was pumped in.

Meanwhile, attempts were being made to deal with the contaminated land around the reactor building. The surrounding area was covered with a protective plastic film (at rates of up to 300,000 m^2 a day) to hold contaminated dust and soils in place.

Once the leaking reactor core had been plugged, the priority was to encase the damaged reactor in a massive protective concrete shell, to prevent any risk of further contamination. Ventilation, cooling and radiation monitoring systems were built in. Three massive concrete factories were commissioned for the task, and over 6,000 metal structures were pieced together on site. Some reports imply that progress in making the shell was slowed down by shortages of cement, but it was still finished by September 1986.

The giant concrete tomb, descending to nearly 30 m beneath ground level and rising as high as a 20 storey building (Figure 3.2), was a considerable technical achievement. It will provide a fitting and permanent monument to the accident (perhaps the world's largest single tombstone!). It is interesting that *Pravda* science writer Vladimir Gubaryev chose to call his play about the Chernobyl disaster (which was staged by the Royal Shakespeare Company in London) *Sarcophagus* (meaning *"stone coffin"*).

47

SHIFTING PRIORITIES

Once the leakage of radiation from the reactor had stopped and the fire was out, more personnel could be moved on site to assist in the massive containment and clean-up operation. Within two weeks of the accident, phase one (*containment*) was well in hand, and phase two (*decontamination*) could begin.

The first hint of a massive decontamination programme over a wide area of the Ukraine had been given in a *Tass* report on 7 May. It described a range of measures being employed, including the regular washing down of all roads in the region by a special fleet of lorries and the establishment of radiation control check points at all major cross-roads. Double radiation checks were being made on all local food produce, and local radio had been instructed to break its earlier silence and carry regular reports on radiation levels in the area.

Decontamination work will continue in and around Chernobyl for a long time to come (see Chapter Eight). The Russian authorities firmly believe that it will be possible to clean up the area (including the town of Pripyat) well enough to allow people to return to a normal existence there. How long that will take is a matter of debate.

The struggle for containment - *physical containment* on site (to prevent melt-down and curtail the spread of radiation), and *diplomatic containment* to limit the spread of international political fallout - had severely stretched the Soviet authorities.

But while the world watched events in Russia, countries downwind from Chernobyl were eagerly monitoring changing wind patterns and changing atmospheric radiation levels. Even more eagerly, they were making hurried contingency plans to cope with their worst fears - high levels of radiation fallout within their national boundaries. We shall look at these in the next chapter.

4
IMMEDIATE PROBLEMS
FOR HUMANS

Many people in the west were concerned about what was happening to those in the area around Chernobyl. Within days of the accident it had become clear that the Soviet authorities were not facing their responsibilities - they were failing to advise their own citizens and foreign nationals in the Ukraine about the threat to their health and failing to advise how health risks might be minimized. The two day delay in announcing the accident to the world only increased the concern.

Indeed the air of Soviet secrecy which shrouded the whole of the first week of the aftermath made matters much worse than they might have been. It also diminished the credibility of the Soviet Union, because there was a growing feeling across Europe that the Russian authorities were acting with serious indifference to international fears and feelings.

The Soviet authorities were sitting on a time-bomb, and they knew it. They and nuclear experts in the west were painfully aware that the damaged reactor at Chernobyl had released a deadly cocktail containing hundreds of radionuclides. The four main ingredients were caesium-137, iodine-131, strontium-90 and carbon-14 (see Chapter Two).

With these dangerous isotopes floating about in the air over and around Chernobyl, it is not surprising that considerable interest was aroused in the west in the casualty reports coming from the Soviet Union. After all, bad news is always good news for the media. But the interest was more deeply rooted, because the radiation cloud was blowing northwards and westwards over Europe (see Chapter Five), and everyone wanted to know just how bad the disaster was. It affected every single person in Europe.

In this chapter, we shall look at the health problems facing those at the Chernobyl power station and others in the area at the time of the accident. The plight of people in Poland, immediately downwind of the area, will also be considered. The accident and its deadly radiation cloud also created problems in western Europe via international trade in contaminated foodstuffs.

DEATH TOLL AND CASUALTY LIST

There was much speculation over how many had died at the power station, and the lack of Soviet news served only to encourage widely variable estimates in the first few days.

Unconfirmed reports from Kiev were talking of up to 300 dead. The Soviet Council of Ministers insisted that only two had died. A Soviet official arriving in Washington on 28 April said " the figures are in tens - under a hundred - including the injured". United Press International (UPI), a US news agency, circulated an alarming report of 80 immediate deaths and a further 2,000 deaths *en route* to hospital, claiming that the dead were being buried in a radioactive waste repository at Pigorov.

The UPI report subsequently proved to be unsubstantiated, and it highlights the problem of separating fact from fiction when observing serious disasters from a distance. The final count was bound to be high (official figures have put it at 30 dead), but doubtless nowhere near as high as the thousands being talked of in the west. The Soviet authorities took great delight in dismissing reports from the *'bourgeoise press'* as alarmist propaganda, malicious in intent and totally without foundation.

Early casualties

Pierce Wright (Science Editor of *The Times*) commented on 30 April that "the controversy about the risks of nuclear (energy) has moved sharply from arguments about statistics and probabilities to numbers of casualties". As the days passed, a steady trickle of statements on the human toll of the accident appeared. The official death toll stayed at two for some time, both fire fighters who died tackling the blaze. One was killed by falling debris, and the other died in an explosion in the refuelling hall (on top of the reactor core).

Nobody from outside the power station was injured in the explosions and fire. Some of the injured were station personnel, but most were in the fire fighting teams which brought the raging fires under control with apparently little thought to their own health. All of the casualties were suffering from radiation poisoning after being exposed to the dangerously high levels of radioactivity pouring out of the damaged reactor. A few were caught in the blast of the explosion, so they suffered severe burns as well.

Around 200 plant workers and firemen with symptoms of radiation sickness (such as nausea and vomiting) were examined by doctors in Pripyat in the first 36 hours. About 200 were admitted to hospital straight away. The 129 worst cases were put on three special flights to Moscow on 27 April, for immediate treatment in special clinics.

Most of the casualties had been exposed to radiation levels estimated at between 1 and 8 grays. Typical doses were around 2 grays, and many had hopefully reversible damage (see Table 2.4). The most seriously injured workers had been exposed to over 5 grays. Most of the radiation exposure was to gamma rays, but doses had accumulated in a variety of ways - some had breathed large doses into their lungs, others had ingested large amounts. Yet others had radiation burns only on one side of their body. The doctors were going to be fully stretched coping with such a number of casulaties and such a wide variety of unusual problems.

American medical assistance

Doctors around the world knew full well that the only way of saving people exposed to lethal doses of radiation - like the 129 critically ill in hospital in Moscow - is to perform bone marrow transplants. Donors with compatible bone marrow have to be found, and some of their marrow transplanted into the patients. The Soviet Union's expertise and resources in this area were known to be limited, and time was also in short supply - the transplants would have to be carried out within two weeks or the individuals could not be saved.

Dr Armand Hammer (multi-millionaire head of Occidental Petroleum and a well-known philanthropist) spotted the opportunity to help, by funding an American team of bone marrow specialists to fly to Moscow. He immediately sent a telegram to Mikhail Gorbachev with the offer of medical and technical assistance. His offer was accepted.

The instant acceptance - at a time when all other foreign offers of help were being coolly ignored by the Soviet leadership (see Chapter Three) - was seen in the west as an indication of just how serious and urgent the problem was. This was accepted as a warning that vast amounts of deadly gamma radiation must have leaked from Chernobyl. It was a bad omen.

The Hammer plan was to sponsor a 'mercy mission' headed by Dr Robert Gale, a leading bone marrow specialist from the University of California. He is a colourful and charismatic figure, with a record of experimental work in the bone marrow transplant field which some see as exciting and pioneering and others see as excited and cavalier. Amongst his many credentials for the job was his role as chairman of the Advisory Committee of the International Bone Marrow Transplant Registry (a consortium of 128 transplant teams in 60 countries). The Hammer team would call upon a pool of 75,000 volunteers in Britain, the

USA and Scandinavia who were available to donate bone marrow in case of emergency.

It was very much a race against time. Gale flew to Moscow late on Friday 2 May, and left immediately for Kiev where he met Soviet doctors and discussed what had to be done. The first task facing the American-Soviet team was triage - the grim process of sorting out the casualties according to priority into three groups. Some victims were already too ill to benefit from surgery - they were doomed to die. Another group, with lesser injuries, might recover without bone marrow transplant, known to be a risky operation suitable only for those in the middle category of people for whom the risk was worth taking.

Nineteen of the people exposed to over 8 grays of radiation were chosen as suitable for transplant operations. Gale planned to use two techniques - the conventional transplanting of bone marrows from donors, and the more experimental transplanting (for which he had won his cavalier reputation) of liver tissues from foetuses. These provide a source of blood cells less likely to be rejected by the most seriously injured than bone marrow cells from donors' blood. Tissue typing, a prerequisite to the operation, proved very difficult on some of the worst injured. Most of the transplants used marrow taken from relatives of the patients, but the international registry was able to deliver matched tissues to Moscow from 15 different countries within three days.

By 18 May 11 of the 35 most critical patients had already died, including six believed to have undergone transplants. Seven of the transplant patients were still alive by early September 1986 - five months after they would almost inevitably have died without the operation, and the skill and dedication of Gale's team.

Despite the high death rate and the massive cost (estimated at around $600,000, met entirely by Hammer), Gale firmly believes that the mission was a success. After all, it saved some lives and it certainly helped to increase medical understanding of how to deal with serious radiation burns. Gale and others, reflecting on the experience after their return to the United States, soberly concluded that it also showed how impossible it would be to cope with the aftermath of a nuclear attack, which would produce many more seriously injured casualties, decrease the potential donor list and destroy medical facilities (and kill doctors and nurses).

The same sponsor also recognized the need to take a longer perspective on the Chernobyl casualty list. Plans were made for a study, funded by the Hammer Foundation, of the long-term health of the 200,000 or so Soviets who were exposed to potentially dangerous levels of radiation.

On 6 June Gale signed a memorandum with Andrei Vovobiev (chief of the Soviet Central Institute for Advanced Medical Studies) setting out terms of reference and agreeing on international co-operation in the follow-up studies.

Gale has recorded what he did, what he saw and what he felt about the Chernobyl accident in his book *Chernobyl - The Final Warning* .

LONGER TERM HEALTH PROSPECTS

The Soviet authorities were later to quote radiation dose rates of a few roentgens (equivalent to a few tens of millisieverts) an hour in the Chernobyl area immediately after the accident. This implies that a large number of people in the area at the time of the accident will have been exposed to (thus accumulated) an entire lifetime's permissible radiation exposure *within the first day* (see Chapter Two). Little wonder, therefore, that although the number of critically injured was small at first, it rose through time as the high dose of radiation to which individuals had been exposed started to take its toll on their skin, bones and organs.

Western experts expected the toll to continue to rise because of the high radiation doses involved. Some clues as to how high these might be were to come from measurements of radiation exposure rates at six sites in western Russia and made available to IAEA since 9 May (Figure 4.1). Dose rates in areas distant from Chernobyl were relatively low, at below 50 microroentgens (equivalent to about 0.0005 millisieverts) an hour. At Oster - 68 km from Kiev - rates remained over 200 microroentgens (around 0.002 millisieverts) an hour until 13 May, and did not fall below 100 microroentgens (around 0.001 millisieverts) an hour until well after 9 June.

Dose rates in all areas must have been much higher straight after the accident, posing serious health risks to those close to the power station. Information released by IAEA suggests that radiation levels in Pripyat were in the region 0.1 to 0.15 millisieverts an hour soon after the accident. They had fallen to around 0.03 millisieverts an hour by 5 May. It is quite likely, therefore, that people in Pripyat received considerable doses of radiation within the first day or two after the accident (before they were evacuated).

One millisievert, which puts people dangerously close to the high risk category threatened with long-term cancers (see Figure 2.3), could easily be accumulated by ordinary people going about their business in Pripyat soon after the accident. Children would have been playing innocently out-of-doors, exposed to direct fallout and in direct contact with contami-

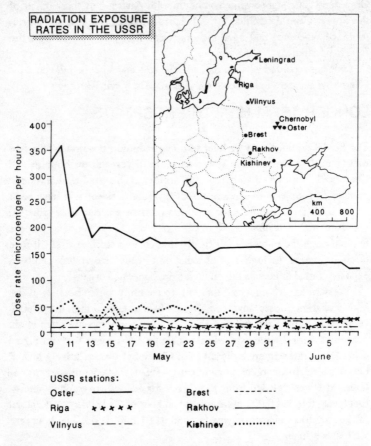

Figure 4.1   *Radiation exposure rates at six sites in the USSR, 9 May to 8 June 1986. Based on information given to IAEA, after Salo (1986)*

nated dust and soil. Children are much more susceptible to radiation damage than adults, and it remains to be seen what the long-term cost of Chernobyl will be to their health .

Doctors in the west identified four main risk zones around the nuclear power station:

(a) within the first 6 km: people were given a 50% chance of surviving without damage to their bone marrow and gastro- intestinal tract. Such damage could kill some of the people within months, and many of the rest within a few decades.

(b) between 6 and about 11 km from the plant few people were likely to die in the short term, but most would experience nausea and other symptoms of radiation poisoning (Table 2.4). Many faced the prospect of developing cancer in the longer term.

(c) within about 100 km of the plant, doctors were anticipating a significant increase in the number of deaths from leukaemia and other forms of cancer within the next 30 years. The greater the distance from the plant, generally the lower the radiation levels and hence the smaller the risk of an individual developing cancer from the fallout.

(d) beyond about 100 km the risks were much smaller.

Several million people were living in the area immediately downwind from Chernobyl (to the north-west) at the time of the accident. Doctors argued that the first cases of leukaemia may start to emerge after 3 to 5 years, and high rates of cancer (perhaps in the order of 1,000 to 2,000 cases per million people) must be expected within that area over the next few decades.

Early estimates suggested that people in Sweden and Eastern Europe were likely to receive radiation doses roughly equivalent to one or two chest X-rays. The long-term medical consequences of such low levels would be extremely difficult to detect.

EVACUATION OF LOCAL PEOPLE

The wind was blowing north at the time of the accident, so the two million people living in Kiev (to the south) were not directly faced with the threat of radiation fallout. Life continued very much as usual there. But people living downwind were threatened.

On the day of the accident there was much indecision and failure to realize how serious the threat really was. That evening (26 April) a local decision was made (by the city Communist Party Committee) to evacuate all people within 10 km of the power station. But the evacuation was delayed for about 36 hours for unspecified organizational problems. Around 25,000 people were moved to safety in Kiev in the 4 hour operation.

The Government Commission, headed by Deputy Premier Boris Shebine, arrived from Moscow on Sunday 27 April to take charge of the disaster. After a few days, during which radiation levels in the area were carefully analysed, they realized the size of the risk.

The danger zone was extended to 30 km, and a further evacuation of people (living between 10 and 30 km from the plant) was ordered. This included residents of Pripyat and Chernobyl, and workers from over a hundred large state farms. The evacuation took place on 2 May (six days after the accident). It was a huge operation which removed 84,000 people in a convoy of 1,100 commandeered buses, stretching nearly 20 km long. Some Soviet reports said it took just over two and a half hours, but others revealed that the process was not completed until 6 May. The evacuation zone became an exclusion zone, manned by Soviet militia units in watchtowers; manned checkpoints were hurriedly thrown across all roads into the zone.

The evacuees were taken initially to Kiev, where they were checked by 230 teams of Soviet doctors and nurses. There are no reports of anyone showing symptoms of acute radiation sickness (Table 2.4).

Most of the 109,000 people were ultimately resettled in new flats and houses in Kiev, but this was no compensation for what they were leaving behind. They left with nothing other than the clothes they were wearing; there was no time to collect belongings, which would be contaminated (thus dangerous) in any case. They had no immediate source of income, and no jobs to go to. The evacuation was so quick that many did not know where their family and friends were. No one had any idea when they might be able to return to their homes and farms - but one thing was certain, their exile was bound to be long-lasting.

No one has been able to explain why the evacuation decision took so long. The delay was critical, because many thousands were exposed to high radiation levels for much longer than was necessary. If they develop cancers in coming decades, much of the blame will rest with those who took so long to order the full evacuation. Local people, living in Pripyat and Chernobyl and in surrounding villages and farms, were given no information about what had happened at the power station and no guidance on how to minimize personal exposure to radiation (the first public announcement by the Ukrainian Ministry of Health was to be given nine days after the accident, when locals had been evacuated any way). Life carried on much as normal for everyone beyond the station.

The Director of Hygiene in Pripyat (Dr Korotkov) was later openly criticized for failing to advise local people what precautions they could and should take - even simple ones like closing windows and keeping children inside. He insisted that he did not have the authority to take such actions. He was later to admit also that he wanted to avoid the spread of panic through local communities. He was subsequently reprimanded by the Soviet authorities.

But the criticisms were not confined to Dr Korotkov. It was clear that the local authorities were ill-prepared for such an accident, and rather helpless in dealing with it. Whilst they did, for example, dispense iodine and put up some local road blocks, they did not issue soldiers working in the area with protective face masks or advise them what to do.

This group of over 200,000 people from the Chernobyl area, who were exposed to high radiation levels for a needlessly long period of time, are the 'high risk' group who stand to pay the biggest long-term cost of the accident (see Chapter Eight). They are the core of the Hammer Foundation's long-term study, who face a real risk of developing cancer (terminal or otherwise) in the future as a direct result of what happened on 26 April 1986.

EXODUS FROM KIEV

Moscow at first tried to paint a picture of ' business as usual' in Kiev, Russia's third largest city (with a population of 2 million), where a report in *Tass* on 7 May described life as " calm, confident and full-blooded ". The reality was somewhat different.

Flight of the westerners

Western technical experts were fairly sure by Wednesday 30 April that radioactive fallout must have contaminated the main water supply reservoir which served Kiev and the water would be unfit to drink. Local fresh milk, meat and vegetables must also be contaminated. The streets and buildings were also likely to be heavily contaminated with radio-active dust and fission products, so that the simple act of breathing would increase radiation exposure.

The British government were concerned about the health of British nationals who were in the area around Chernobyl. It emerged that there were 69 students and three teachers in the middle of a three month study visit to the Kiev State Pedagogic Institute for Foreign Languages, and a further 28 students and two British Council staff in Minsk (about 300 km away). Their lessons continued as usual until Tuesday 29 April, although by then news of the accident was reaching them via radio broadcasts by the BBC World Service and Voice of America.

Hurried arrangements were made for the evacuation of both groups by the British Embassy in Moscow. They travelled on overnight trains to Moscow on the Wednesday, and flew back to London on the Thursday morning (May Day). Each student was first screened on arrival at London airport, where the level of iodine-131 in their thyroid was measured by geiger-counter. They were then given full medical checks in Oxford.

They had been given fresh overalls and track-suits to change into before they left Moscow; their own clothes were packed in special plastic bags, returned to the UK with them and checked for radiation levels on arrival. Some of the party's clothing was contaminated, but they had not ingested enough radioactive material to give a real risk of thyroid cancer.

Japanese tourists who had been in western Russia at the time of the accident had their luggage checked when they returned home on 11 May. Some of the luggage had radiation levels over 100 times higher than normal, and several items were placed in custody by the Japanese government on health grounds.

Foreign companies also advised their workers to leave the area. A hundred Finnish construction workers and a further 50 tourists were brought home from the Kiev area on 30 April. A special flight was also chartered to fly home relatives of Austrian steel specialists working at Shlobin (about 150 km west of Chernobyl). Fourteen British engineers who had been working about 80 km south of Kiev flew home to London on 2 May.

The Soviet Ministry of Foreign Affairs had said on 29 April that there was no reason for tourists not to visit Kiev and nearby cities. That same day the UK Foreign Office warned Britons not to travel to the western Soviet Union unless their journey was absolutely necessary. This warning was extended the next day to cover north-east Poland. Members of the London Festival Ballet voted unanimously to cancel their planned three-week spring tour of the Soviet Union.

On 30 April Americans were advised by their government not to travel to the Kiev area, and Finns were officially advised not to travel to the southern Soviet Union, or to eastern bloc countries near the Ukraine. A meeting of the EEC foreign ministers was to advise travellers two weeks later (on 12 May) to avoid Kiev, the western Ukraine, Minsk and Byelorussia, Lithuania and north-east Poland.

A further measure of international concern over the risk of radiation contamination in the Kiev area came on 5 May when Britain, the United States, Switzerland, Belgium, West Germany, Yugoslavia and Romania withdrew their teams from the prestigious 39th annual Peace Amateur Cycling Race scheduled to start near Kiev the following day. The Soviet authorities insisted in running the race, even with its much depleted international field, to demonstrate its confidence in the official diagnosis that there was no risk to health in the Kiev area.

Migration of locals

There had been little apparent unease amongst the Soviet people in Kiev, even after they were told (on 30 April) that they could check their personal radiation levels at special centres set up at poly-clinics in the city. The unease began on 2 May, when local health authorities issued warnings on television *not* to eat leaf vegetables (especially lettuce) and *not* to let children play outside for more than short periods.

A change in wind direction then started to blow radiation fallout southwards from Chernobyl, directly over the city and surrounding area. Local anxiety rose considerably when orders were given to wash down the interiors of all flats with cold water and avoid swimming in outdoor reservoirs. The advice to leave the area, given to foreign nationals several days earlier, had been well founded! It was to emerge later that the British Embassy had monitored radiation levels in Kiev after the accident, and found dose rates as high as 0.03 millisieverts an hour during early May (70 hours exposure would equal a normal year's exposure to background radiation (Table 2.1)).

By 7 May new measures were introduced which made it essential for all people leaving the Kiev area to undergo radiation checks. The selling of ice cream, cakes and drinks on the streets was banned. Police conducted spot checks for radiation, and water trucks hosed down streets to wash away radioactive dust. Children in the city were also ordered to remain indoors during school breaks. Most western observers believed that the unease had by now turned to panic, but the Soviet authorities strenuously (and predictably) denied this.

It is not entirely clear when the human exodus began, but by 8 May planes and trains arriving in Moscow from Kiev were full of local women and children (many of them unaccompanied). What started as a slow seepage of voluntary departures had become a full and centrally-organized torrent of evacuees by the end of the week (Friday 9 May) when it was officially announced that all children between 6 and 13, and all breast-feeding mothers, were being sent away from the area for the summer.

The evacuation was announced as a precautionary measure, and the Soviet authorities were keen to play down the prospect of mass public panic (in the Ukraine and further afield in Russia) by stressing that all they had done was to start the customary summer holiday for 250,000 local schoolchildren - who normally go away for the summer in any case - two weeks early. Scheduled trains from Kiev to Moscow were packed, and extra trains were laid on to cope with the flood.

Official reports that 'all is well' in Kiev did not reflect what was happening on the ground. Emergency wells were being dug, in case local reservoirs became contaminated. Officials were advising locals not to play soccer or volleyball on river beaches, to avoid the spread of radioactive dust into the rivers themselves. There were genuine signs of anxiety!

Conflicting views

May Day had fallen conveniently in the middle of the immediate aftermath of Chernobyl, and it provided a much-welcomed opportunity to deflect attention away from the disaster. Soviet eyes were deflected; western ones most certainly were not! Russian television and newspaper coverage of the day's festivities made no mention of Chernobyl. In Moscow's Red Square, Mikhail Gorbachev was seen presiding over the traditional May Day parade - a carnival of balloons, sunshine and unashamed ideology. Coverage of the parade in Kiev centred on smiling groups of local girls in traditional costumes. "*No cause for concern* " was the defiant message.

But behind the festive scenes there was growing anxiety about the safety of people, even in Moscow, given the risk of contaminated food and water supplies. Feelings were running high amongst Moscow's 8,500 foreign community who were being advised - as they watched the Red Square celebrations of all that is good in Russia - not to touch milk products over the next 30 days, and to be careful about buying contaminated meat, vegetables and fish. The Soviet government was clearly still trying to contain the political fallout from Chernobyl and restore its flagging diplomatic credibility when the Council of Ministers reported (through *Tass* on 1 May) that no foreign citizens had been affected by the disaster. Foreign citizens were, however, trying to take as few risks as possible.

Western experts also expressed anxiety over the health of people in Minsk, over 300 km north-west of Chernobyl and right in the path of the radiation cloud which eventually reached Sweden (see Chapter One). Minsk, the capital of Byelorussia with a population of around 1.25 million, must have received very high levels of radiation fallout. But, except for the evacuation of the British students, the city showed a disarming sense of normality. Local residents were given no official advice on how to minimize personal risk (such as not drinking local milk), and western correspondents found a population totally unconvinced (certainly by 4 May) that they faced any real danger whatsoever. This lack of concern, especially in one of eastern Russia's major food processing centres was itself a matter of concern to observers beyond the Iron Curtain.

POLAND - UNDER THE SHADOW

By providence of location, Poland lay immediately downwind of Chernobyl at the time of the accident - it was in the direct shadow. Radiation levels in the air were high, and the cloud hung overhead for many days. But Polish problems were compounded because it seems that Moscow did not even tell its communist neighbour about the accident until it was too late to take proper (and effective) remedial and preventive action to safeguard humans. The country was plunged into a serious crisis.

The first evidence of high atmospheric radiation levels over Poland emerged early on the morning of Monday 28 April, while Swedish scientists at Forsmark - nearly 700 km away to the north - were discovering the same thing (Chapter One). The Polish radiation monitoring service (SPSD) runs 140 permanent stations around the country. Its monitoring station in the village of Mikolajki, in the Polish lake district (some 500 km west of Chernobyl), was giving readings up to 500 times the normal background radiation level that Monday morning. By mid-afternoon radiation levels in the air (gross beta activity) had risen even higher, to around 550 becquerels per cubic metre.

The next day above-average radiation levels were being recorded on over 200 civil and military geiger counters throughout north-east Poland. By now the cloud covered a large area (stretching from Suwalki in the far north-east down as far as the cities of Olsztyn (150 km north of Warsaw) and Bialystok (about 170 km to the north-east, close to the border with Byelorussia). Warsaw, the capital city, was not yet at risk but it would be if winds were to blow the cloud further south.

Crisis team and swift reactions

The Polish authorities wasted no time in setting up a top-level joint civil-military crisis team to draw up emergency plans. The Governmental Committee for Assessment of Nuclear Radiation and Prophylactic Measures, headed by the Deputy Prime Minister (Mr Zbigniew Szalajda), was formed early on the Tuesday morning 29 April). It held its first meeting straight away, and issued a communique the same day which advised that:

(a) hospitals in the area were being put on standby alert
(b) (uncontaminated) food reserves were being mobilized
(c) people were advised by the Ministry of Health not to buy or

   drink milk from north-east Poland
(d) no restrictions were expected on the sale of foodstuffs.

The overall strategy was to be prepared but play down the problem, in order to avoid public unrest and widespread panic. The general belief in Poland was that the radiation cloud posed no real threat ... so long as it kept moving overhead. The risk of fallout would inevitably increase sharply if rains washed the radioactive particles out of the cloud, but the prevailing hot weather and uncustomary lack of rain at this time of the year were welcomed as hopeful signs. Polish fingers were kept tightly crossed during the early part of that week!

   Radiation levels around Bialystok started to fall after the Monday, and the cloud began to drift westwards over north-east Poland. On the Tuesday (29 April) the crisis team made a series of recommendations designed to minimize health risks. They advised that consumption of milk from cows fed with green fodder be banned, and that milk should only be used after industrial processing. Cows should be kept in barns and not fed with green fodder, they said. The distribution of pre-accident powdered milk for children up to three years old was also recommended, and a banning of green vegetables, meat and fish with high radiation levels (above 5,000 becquerels per kg) was proposed.

Iodine prophylactic

The crisis team also recommended that stable iodine (used as a prophylactic, or preventive medicine, and referred to locally as *'anti-fallout medicine'*) be distributed to all children under ten years of age. The Ministry of Health and Social Welfare took on responsibility for distributing the iodine. They decided to administer a single dose in solution, and Poland's 3,348 pharmacies set to straight away to prepare and distribute the material through hospitals, schools and nurseries as well as dispensing it themselves.

   Work in administering the solution began on the Tuesday evening in the Bialystok region. The next day iodine was being administered throughout the country. An estimated 10 million children (98% of the under-16s) received their dose of stable iodine from doctors who visited every creche and school in Poland. Several million adults took iodine by choice (although it was not recommended by the authorities). The pharmacies were hard pushed to produce the iodine solution so quickly and in such vast quantities, and rumours spread quickly of grossly inflated black market prices for the potent liquid.

But the liquid iodine is not guaranteed to work. It must be given in the right doses at the right time if it is to successfully counteract the effects of radioactivity in humans. Moreover, if too much iodine solution is swallowed, this can lead to iodine poisoning as serious as that which might come from exposure to the iodine-131 fallout. It was very much a double-edged sword. Health experts in the west quickly pointed out that most clinics did not know anything about local radiation levels when they gave the iodine solution, so that the doses given must have been guessed at in a rather arbitrary manner. Whether people were helped or hindered by the iodine solution administered under such conditions will perhaps never be known.

Washout in Warsaw

Dry weather persisted until the worst of all possibilities occurred on the Wednesday afternoon (30 April) - a heavy rain fell over Warsaw just as the radiation cloud was overhead. The problem quite literally landed on the ground, in the form of highly radioactive dust and particles.

Residents were advised to take a shower several times a day, to reduce the risk of serious contamination. Children were advised to wash their hair and change their clothes each day - an inconvenience of the highest order of magnitude! Parents were advised (unofficially) to keep their children indoors as much as possible - an inconvenience of equal g ravity for the hapless parents!

The western community in Warsaw was advised by the British Embassy on 30 April that they could minimize the personal risks by taking several simple steps - avoid drinking or using fresh milk, avoid drinking tap water, wash children and clothes regularly. The same day the Embassy was making plans to evacuate pregnant wives and young children of its staff from the ' *high risk area* ' around Warsaw and the north-east, and advised British companies operating in Poland to do the same. Foreign Office officials in London also advised Britons not to travel to north-east Poland until further notice.

Risk and reactions

By mid-week there was a dark cloud of gloom hanging ominously over Poland. Western experts anticipated radiation levels up to 20 times higher than normal over eastern Poland that day. But they also predicted that the health risk to most humans would be minimal. If the radiation cloud stayed overhead for ten days, they estimated, people would still only receive around 0.7 millisieverts of radiation (about a third of the average person's exposure to natural background radiation). The only groups at risk would be small babies and pregnant women.

Polish experts were less optimistic. Professor Zbiegniew Jaworowski, head of the country's Central Laboratory for Radiological Protection, forecast an increase in the order of a few percent in the incidence of thyroid cancer, with a total increase in radiation-induced cancer deaths in Poland of between 200 and 500 over the next 30 years. He also feared a marked increase in bone diseases and leukaemia, and a significant risk that babies about to be born would suffer from malfunctioning thyroid glands.

By the Friday (2 May) the Polish crisis team were convinced that levels of radioactivity in the air were falling, but levels in soil and water were still high. Many were hoping for a long spell of heavy rain which would wash the radioactive dust and particles down into the lower soil. But public anxiety continued - the sale of milk from grass-fed cows was formally banned, other dairy products were voluntarily removed from the shelves in shops, and iodine solution had by now been dispensed to all children under 16. Concern among the foreign community also c ontinued to rise, and they were voting with their feet - by now all scheduled flights to the west were fully booked.

The radiation situation remained fairly unchanged over the weekend - levels in the air were falling whilst levels on the ground remained relatively high. New contingency measures were introduced as the crisis grew deeper. Parents were warned not to let their children play on grass or in sand pits, for fear of contamination.

Considerable amounts of radiation had by now fallen over the grazing land in north-east Poland, and fears were growing that milk and milk-products were being contaminated. Milk sales had been restricted throughout the country since the start of the crisis. By 3 May the situation was more serious, and many shops were instructed not to sell cottage cheese or milk over the following two weeks. A Government communique issued on the 5th May announced that the sale of milk from grass-fed cows was being banned, that contaminated milk was being processed for industrial use, and that rationing of powdered milk was being introduced (to make sure that enough was available for infants less than 6 months old).

The government hoped that the ban on milk from grass-fed cows would be short-lived. Western observers believed that the main reason was official anxiety that some dairy farmers were mixing contaminated milk from grass-fed cows with clean milk from cows fed indoors (to reduce their loss of income, which by now was getting serious). It was thought that the Polish authorities were not convinced that the grass was no longer seriously contaminated.

On 8 May the crisis team recommended that children, pregnant women and women who were still breast-feeding their babies should not eat vegetables like lettuce, spinach and sorrel because of the risk of contamination.

Radiation levels in Poland

Figures published by the crisis team showed that, in the week after the accident, iodine-131 levels in milk from north-east Poland had ranged from 200 to 1,720 becquerels per litre. Much of this was well above the internationally agreed emergency level of 1,000 becquerels per litre for children (Chapter Two). Iodine levels in milk across the rest of the country ranged from 0 to 600 becquerels per litre - high but generally not too serious.

The crisis team also published figures on the levels of radiation which people in Poland had received during that first week (up to 2 May). The average accumulated dose was put at 0.25 millisieverts, about a twentieth of the 5 millisieverts a year taken as an acceptable upper limit for the public (Chapter Two). However this was a national average, and it masked out strong regional variations reflecting the distribution of fallout across the country (highest in the north-east).

US scientists at the Lawrence Livermore National Laboratory later predicted what doses of iodine-131 adults across central Europe were likely to have accumulated from the Chernobyl accident. The results of their simulation, covering the period 26 April to 1 May, are shown in Figure 4.2. These estimates suggest that people across Poland received much larger radiation doses than the figures given by the Polish authorities. Those living in the far north-east might have received over 10 millisieverts of iodine-131 alone, and people in central Poland (including Warsaw) were probably exposed to more than 1 millisievert, during that five day period alone. Iodine-131 has a half-life of 8.1 days (Table 2.3), so most of the damage will have been done within the first week.

BAN ON FOOD EXPORTS FROM RUSSIA

Governments in the west could only watch as the radiation cloud floated over Europe, leaving a deadly trail of highly dangerous dust and fission

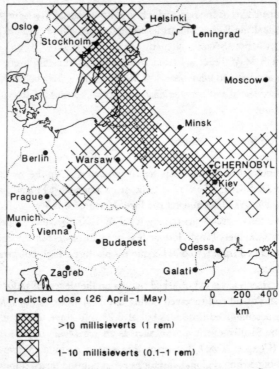

Figure 4.2   Predicted doses of iodine-131 in the USSR and surrounding countries. Based on a simulation of the integrated dose to adult thyroids accumulated from 26 April to 1 May 1986, by Lawrence Livermore National Laboratory. After Levi (1986)

products behind it (Chapter Five). They had no control over where the cloud moved, what it carried or how much it deposited. But they could at least minimize public health risk at home by taking a responsible attitude to the import of contaminated foodstuffs from the area closest to Chernobyl, which had doubtless received the highest radiation levels.

EEC diplomats in Brussels first proposed a temporary ban on the import of food from the Soviet bloc countries, on Monday 5 May. It was stressed that the proposed ban would be a " temporary but indefinite" contingency measure against contamination, which would " make it possible to remove any restrictions on the free flow of trade throughout the Community while at the same time protecting the health of the public".

The plan was to ban the import of some fruit and vegetables, milk, fresh meat, animals for slaughter, game and freshwater fish, from six countries within 1,000 km of Chernobyl (Bulgaria, Hungary, Poland, Romania, Czechoslovakia and the Soviet Union). The intention was to exclude Austria and Yugoslavia from the ban because, although contamination might have been high, at least the EEC was being kept informed about what was happening in those countries. Senior EEC diplomats were later to extend the range of products to be banned, and to extend the ban to cover Yugoslavia. Serious thought was also given to including Austria, East Germany and Albania.

EEC trade and consumer affairs ministers met on 6 May to consider this proposal. They were unable to agree on the controversial plan and referred it to tactical experts. The following day it was discussed by the group of senior EEC diplomats who are empowered to take executive decisions when the Council of Ministers is not in session (as then), who referred it to the next meeting of EEC foreign ministers, scheduled for the following Monday (12 May).

The European Commission did, however, take some action on 8 May, imposing an immediate ban on the import of all fresh meat and live cattle and pigs from eastern Europe. This was exercising its fullest powers, because the Commission needs approval of the Council of Ministers to ban other foodstuffs. It would also have a significant economic effect, because these items account for over two-thirds of the food imported by the EEC from the eastern bloc countries affected by the ban.

The main point of disagreement between the 12 member governments was over the proposed scale to be used in measuring radiation in foodstuffs traded within the Community. France insisted that the *'Maximum Radiation Tolerance Levels'* - a scale based on the maximum permissible intake of radiation for babies in milk, fruit and vegetables - should be used, but other countries objected to this as restrictive. Italy argued that the tolerance levels for fruit and vegetables were too low in comparison with those for milk.

The proposed ban was greeted with predictable dismay in the Soviet Union. On 8 May Moscow attacked it as contravening the stated aims of trade, economic and scientific accord. Mr Vladimir Lomeiko, Russia's chief Foreign Ministry spokesman, argued on 12 May that the plan was discriminatory and "without a single justification". Poland joined the attack on the plan on 11 May, dismissing it as an "expansion of political restrictions ... a dishonest trick of competition" inspired more by protectionism than by genuine concern about human health.

The slow passage of the proposed ban through the EEC bureaucratic maze made many countries anxious that the delays would allow contaminated foodstuffs to arrive, and be sold, on their own home markets. Some took unilateral action. For example Italy had already banned the import of foodstuffs from eastern Europe and from fellow EEC states on 7 May, and Spain followed the next day with a ban on Russian food.

The British government saw the proposed ban as "unnecessarily restrictive", and was only persuaded to join the call for it on 8 May. The same day Britain imposed its own temporary ban on the import of Soviet foodstuffs, scheduled to be reviewed on 20 May but to last until at least the end of the month. British Members of Parliament, discussing the proposed ban in Parliament, recognized a serious loophole in it. This stemmed from existing trade agreements between East and West Germany which would make it impossible to determine whether foodstuffs (such as pork, bacon and some vegetables) coming from West Germany might have originated in (contaminated) eastern Europe.

By 9 May member governments within the EEC had still failed to reach agreement on the proposed ban - two weeks after the Chernobyl accident and five days after the proposal had first been suggested. The following day France imposed its own ban on the import of all food products from eastern Europe, except East Germany. By 12 May eleven of the twelve member states of the EEC were operating their own bans against food imports from eastern Europe, in line with the EEC proposals. These bans were to remain in force throughout May.

Concern about the possibility of importing contaminated food was not restricted to continental Europe. On 11 May the US Government added its weight to the growing international anxiety, and directed inspectors of ports and airports throughout the country to closely monitor food (especially soft cheese, fresh fruits and fresh meat) imported from 11 countries that might have been contaminated by fallout from Chernobyl. The hit list included Austria, Czechoslovakia, Denmark, East Germany, Finland, Hungary, Norway, Poland, Sweden, Japan and the Soviet Union. It ordered the immediate ban of all imports from any country from which contaminated items were being imported, until further notice.

These food bans were well intended, and designed to minimize the threat to human health. But the experts were divided over whether they would really be needed, or really be effective. When radiation experts from 11 countries were called together by the World Health Organization for a high-level meeting in Copenhagen on 7 May (see Chapter Five), the general feeling was that by then restrictions on food imports could not generally be justified on health grounds, and blanket bans on the import of fresh food from countries in eastern Europe (except for the Soviet Union) were no longer necessary.

PROBLEMS NEAR AND FAR

People thoughout Europe sensed the danger from Chernobyl. Here was a nuclear power station, badly damaged, and pouring deadly radioactive material uncontrollably out into the air for ten days. The wind was clearly blowing it over a wide area. Was anywhere safe from the threat?

The official Soviet report on the accident (see Chapter Seven) was later to estimate that the initial explosion and fire released around 444,000 million million becquerels (12 million curies) of radioactivity into the environment during the first 24 hours, and a further 1,400,000 million million becquerels (38 million curies) before the reactor was finally sealed. It was in all senses an unprecedented accident - in terms of amounts released, area covered, damage done and fears raised. There was little to smile about!

Some estimates show that up to half of the ejected material may have been deposited within as little as 30 km from the power station. This was little consolation for the workers and firemen exposed to dangerously high radiation doses on site, or to the 100,000 or so people who were eventually evacuated from within this 30 km danger zone. But it did give a glimmer of hope to people outside the immediate area. Poland was too close to the radiation source for comfort, as we have seen. Elsewhere, all eyes were on the radiation cloud and its likely paths across Europe (Chapter Five).

5

UNDER THE SHADOW - THE RADIATION CLOUD

The world awoke on Tuesday 28 April to growing concern about the drift of radioactive fallout from the accident. By now it was apparent that massive amounts of radiation had leaked from Chernobyl, much of it material with a dangerously long half-life. There was mounting unease at the prospect of the radiation cloud blowing westwards across Europe. Most countries were on the alert in this uniquely sinister game of Russian Roulette. Everyone was, quite literally, at the mercy of weather patterns.

In this chapter we shall examine how the radiation cloud drifted across Europe during the first ten days or so, and how the fallout was deposited. This long shadow of Chernobyl, cast over many countries, was to be a source of widespread concern and anger beyond the Soviet Union. Nowhere appeared to be safe - radiation from the accident was ultimately to be detected as far away as Japan, Canada and the eastern United States.

INVISIBLE DRIFTING MENACE

Western scientists initially believed that the reactor fire would not generate enough heat to lift the plume of fission products high into the atmosphere. They speculated that most of the radioactivity would fall within a radius of about 100 km around the plant. But Tom Wilkie was to report in *New Scientist* on 15 May that "the force of the blast carried much of the initial release of radioactivity high into the air, where it was eventually blown north towards Scandinavia. Because of the force of the explosion, it seems likely that there was little local fallout of radioactivity in the early stages of the accident."

The initial release of radioactive material was propelled high into the atmosphere by the explosion and the immense heat of the graphite fire. It could be blown long distances and quickly by upper air winds. But material (mainly radioactive particles and gases) continued to seep from the damaged reactor for 12 days, before it was eventually sealed (Chapter Three). Much of this remained lower in the atmosphere, but it could also be spread far and wide by prevailing winds.

Many news reports at the time spoke of a *'radiation cloud'*, but this suggests a distinct plume of radioactive material floating about in the sky and visible to the naked eye. The material was invisible and detectable only with sensitive monitoring equipment. But we can use the term *'cloud'* for simplicity, accepting this qualifier.

That the cloud had no respect for national frontiers was beyond doubt. All that each country in its path could do was hope that a change in wind speed or direction would make it pass overhead as quickly as possible and migrate into someone else's territory. They could also hope that it would not rain while the cloud passed over their territory, because this would wash the radioactive particles down to earth.

Nordic activity

Radiation monitoring stations across Europe were on constant vigil over the threat of receiving the unwanted invisible import. Throughout Scandinavia levels of radiation were falling by 28 April (ie within three days), but they were still much higher than usual. Sweden's Institute for Radiation Protection revealed that radiation levels over the country had fallen by half overnight, but were still between ten and a hundred times higher than normal. Measurements taken over the following two weeks were to show a progressive fall in caesium-137 and iodine-131 in the air over Stockholm (Figure 5.1), interrupted by a rise on 9 May when the radiation cloud was blown once again over Scandinavia (Figure 6.1).

At this early stage radiation levels were still extremely high. The radiation cloud over Sweden had also spread over a larger area by now. High radiation was being recorded down the east coast from Gavle (north of Forsmark) to Nykoping (south of Stockholm). The Baltic island of Gotland was also under its shadow.

Similar patterns were reported throughout Scandinavia - radiation levels which were falling, but still much higher than normal. In Norway the Institute of Radiological Hygiene continued to measure high levels of radiation, although levels were falling in the east. Danish scientists reported radiation levels up to five times the norm. Radiation levels were still high in Finland; highest at Kajaani in central Finland, near the Soviet border.

A fairly strong north wind blew the radioactive cloud south during Tuesday 28 April. By now anxiety was mounting rapidly all around

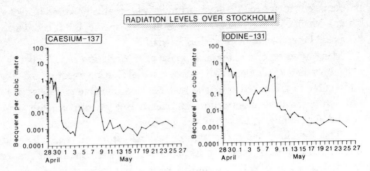

Figure 5.1   Radiation levels over Stockholm between 28 April and 27 May 1986, showing variations in caesium-137 and iodine-131. After Jensen & Lindhe (1986)

Europe, as changing weather patterns threatened to bring fallout over central Europe (Figure 5.2). The continued lack of information about what had happened at Chernobyl, what was leaking out and how dangerous it might be (see Chapter Three) served to increase tension even further.

It was becoming clear that people across Europe wanted and needed guidance about what the risks were. In the first few days after the accident was announced stories about the accident and the drift of radiation monopolized the news around the free world. Speculation and inference were often passed on as fact, few people really understood what was happening, and many people were starting to lose a lot of sleep over Chernobyl.

The International Atomic Energy Agency (IAEA) came to the rescue on 30 April, when it announced that radiation levels appeared to pose no real threat to humans, certainly outside Russia. They predicted that most people in Europe were likely to face radiation doses only a few percent higher than the average annual natural dose. Whilst this did put some minds at rest around Europe, many countries chose to err on the cautious side and press ahead with making their own contingency plans. These seemed clearly justified by the belief (confirmed by IAEA) that some people might ingest large amounts of radioactivity via contaminated foodstuffs.

Contingency plans

By the end of April 1986 countries had neatly divided themselves into three groups. One group (Sweden, Denmark, Poland and Finland) was already exposed to radiation and thus engaged in the *'front line'* of the battle. The second group (all other countries in western Europe) were in the vanguard, and it was only a matter of time before they were to be drawn directly into this particular theatre of war. Group three (countries like the USA and Japan) - the envy of the rest - were fortunate in being located long distances from Chernobyl; they could remain detached observers.

The game-plan was unfolding fast, and most countries in the first two groups were receiving radiation from Chernobyl by 2 May (Table 5.1, Figure 5.2).

Table 5.1 Dates of first measurement of increased radiation levels

The dates on which measurements of increased radiation levels were made, as reported to the International Atomic Energy Agency -

Sunday 27 April	Finland, Sweden, Denmark, Poland
Tuesday 29 April	Norway, East Germany, Hungary, Austria, Yugoslavia, Italy
Wednesday 30 April	West Germany, Switzerland, Turkey
Thursday 1 May	France
Friday 2 May	Greece, Netherlands, Belgium, UK

SOURCE: summarized from Salo (1986)

The *'front line'* countries reacted to the radiation threat in a variety of ways. By Wednesday 30 April patterns and levels of radiation over eastern and central Europe had started to change (Figure 5.2). Radiation levels over Finland were by then declining as air currents pushed the cloud of radiation away to the south-west. The cloud still hung ominously over Sweden, posing a serious threat of contamination over a wide area. Officials in West Germany were growing concerned that the easterly wind was blowing the radiation towards them (Figure 5.2).

Anxiety was also mounting in central Europe (Figure 5.2), as changes in wind direction carried the cloud further south and isolated rain storms brought the radioactive material down to earth. Radiation levels in eastern Switzerland were up to ten times higher than normal by 30 April, and up to 17 times the norm in Carinthia, Austria's most southerly state.

By now plans were being set in motion to cope with the fallout threat. Sweden was quickest off the mark and best prepared. Officials at the National Institute of Radiation Protection estimated that people in Sweden had already (by 30 April) been exposed to similar amounts of

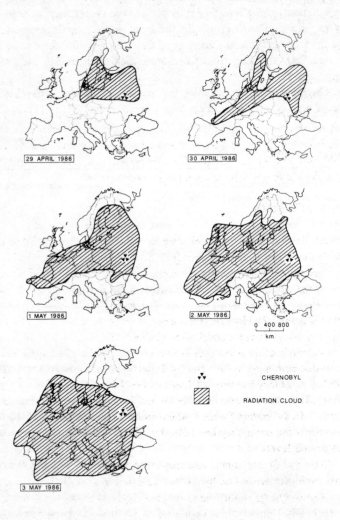

Figure 5.2   Spread of the radiation cloud across Europe, from 29 April to 3 May 1986. After Loprieno (1986)

radiation as people affected by the Three Mile Island accident (see Chapter One). The government recognized the need to minimize the risk of further human exposure. It imposed a ban on the import of food from the Soviet Union, days before other countries woke up to the same possibility. People living along the east coast between Uppsala and Umea were advised not to drink or use rainwater, when radiation levels there were found to be rising. By 30 April the Swedish Ministry of Agriculture had detected iodine-131 in some milk samples, but not in concentrations high enough to justify restricting the sale of fresh milk.

Few other countries adopted formal measures at home (such as the restriction of movements on food and drink) at this early stage, because radiation levels were still way below anything critical. But whilst people at home might have been safe, it was clear that nationals who were working or on holiday in the Ukraine faced the real risk of direct exposure from radiation fallout and indirect exposure via contaminated food and drink (milk and water). A flight of the westerners (see Chapter Four) had begun days after the high radiation levels came to light.

Cloud drifts south

Towards the end of the first week after the accident, the focus of attention outside the Soviet Union switched from Scandinavia towards Poland and the other eastern bloc countries in Europe. Prevailing winds had started to blow the cloud south from Scandinavia on the Thursday (1 May), and overnight winds had carried radiation over much of southern Europe by 2 May (Figure 5.2). High radiation levels were by now being reported in Italy, Yugoslavia, Greece and Turkey (Table 5.1).

Scandinavia could at last start to heave a sigh of relief. Radiation levels in Sweden continued to fall and by Thursday they were around a fifth (20%) of what they had been 24 hours earlier. Levels of caesium-137 and iodine-131 in the air over Stockholm had by now fallen considerably (Figure 5.1). In eastern Sweden, where radiation levels had been highest, at ten times the norm, they had fallen by Thursday to just double normal background levels.

Countries further south were inevitably somewhat displeased to receive the legacy passed on by the Nordic countries. Radiation levels in West Germany rose slightly early on Thursday 1 May, then fell steadily. Levels in East Germany were by now up to 100 times higher than normal, but this was not considered a direct threat to health. The cloud arrived over the Netherlands overnight, and the level of iodine-131 in the air over Groningen rose sharply from zero to peak just below 20 becquerels per cubic metre by early afternoon on 2 May. Within 24 hours it was down to a quarter of this level, and it was below 1 becquerel per cubic metre by the morning of 4 May.

In Switzerland radiation levels continued to rise up to ten times the norm on the Thursday (1 May), and they began to fall again the next day. Despite assurances that this posed no threat to health, there was some panic-buying of food in Switzerland - mainly evaporated milk, bottled mineral water and bags of frozen vegetables (picked and processed before 25 April). A short-term ban (lifted on 8 May) was imposed on the sale of milk from goats, sheep and cows (pasteurized).

By early May the cloud had started to cast its shadow over southern Europe as strong northerly winds pushed the radiation cloud further south (Figure 5.2). High radiation levels were recorded in Yugoslavia, Romania and Bulgaria (south of the Balkans). Czechoslovakia remained curiously silent, and issued no information on whether any rise in radiation had been detected.

Over the weekend (3-4 May) the cloud continued to move south allowing a partial recovery in Scandinavia (Figure 5.2). Officials in Sweden were already anticipating that radiation levels could have fallen back to normal within a week (Figure 5.1). But to minimize the risk of radiation poisoning to humans (via milk), farmers were instructed to keep milk-producing cows under cover and not to feed them on fresh (possibly contaminated) grass until further notice.

In central Europe the improvement continued. By Sunday (4 May) levels of radiation across most of West Germany had more or less returned to normal, except in the south. As a precautionary measure people in the state of North Rhine-Westphalia were advised not to let their children play in sandboxes outside, not to touch the ground with anything but their feet, and to protect themselves from rain. Precautionary checks were also being made throughout West Germany on radiation levels in all fresh milk and vegetables, and on lorries from the eastern bloc which were entering the country at border crossing points. Similar checks were being made on aircraft arriving in Luxembourg from eastern Europe, and people in that country were advised to stay inside during rain showers.

A measure of the significance of Chernobyl fallout was given by Mr Otto Huber, president of the Swiss Radioactivity Monitoring Commission, who revealed that by 5 May levels of fallout had already exceeded that from all previous atmospheric testing of nuclear weapons around the world added together. There were further fears, because even though radiation levels in the air might be falling, much of the radioactive material in the cloud had by now fallen (or been washed) to the ground. Radiation levels in grass (Figure 5.3) and crops, in and on the soil, and in and on buildings and structures would remain high for some time to come, and much of this material could affect humans (see Figure 2.2). The

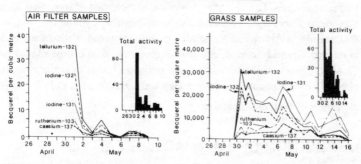

Figure 5.3   Radiation levels at Siebersdorf, Austria. Based on measurements taken at the IAEA laboratories from 26 April to 10 May 1986. After Schelenz & Abdel-Rassoul (1986)

problem was by no means solved.

In southern Europe the situation remained grave by 3 May (Figure 5.2), and it appeared to be deteriorating. Romania declared a state of alert after a significant rise in radiation levels was measured, with people advised to stay indoors until further notice and not to drink water from wells or eat fresh vegetables. Bulgarian radio curtly announced that radiation levels over the country posed no danger to humans. Turkish radio announced that radiation levels in the soil and rainwater were high near the border with Bulgaria, and advised people not to touch rainwater in that area. The northerly breeze over the weekend brought the fallout over Greek soil (Figure 5.2). When radiation levels were found to be four times higher than normal Greeks were advised to wash fruit and vegetables thoroughly before eating them, as a precaution.

The cloud had by now also spread over south-west Europe (Figure 5.2). In Spain, rain which fell on the Sunday (4 May) had above-normal levels of radiation, but no precautionary steps were thought necessary. Although high radiation levels over France were being reported to IAEA by 1 May, the French government kept noticeably silent about radiation problems until 12 May. Critics argue that the two week delay in making any formal announcement about levels of radiation over France reflects a nervousness about how to handle problem created in the aftermath of Chernobyl, borne of the country's abiding faith in the safety of nuclear power.

A country which gets around two-thirds of its electricity (64.8% in 1985) from nuclear power will clearly be eager to play down any prospect of public fear and anxiety of radiation pollution. The French Radiation Service initially announced that no significant radiation increases had been detected over France. But figures released on 13 May by Pierre Pellerin, Director of the Service, showed that France *was* in the shadow of Chernobyl. Radiation levels had been up by a factor of up to 400, mainly in Alsace and the south-east, but they had since fallen back to normal. Spinach from Alsace was temporarily taken off the market on 15 May. French officials insisted that radiation levels over the country posed no risk to people's health, so they had felt little need to fully notify the public of their measurements.

DEPOSITION OF RADIOACTIVE FALLOUT

The real problem with the radiation cloud which passed over Europe was what it was likely to leave behind. Radiation levels in the air, certainly outside the stricken area within the USSR, were not high enough to put human health directly at risk. But there was a clear indirect risk from contamination of soils, crops, water supplies (which could be eaten and drunk) and from deposition of radioactive dust (which could be inhaled) over a wide area.

Measurements

Measurements of levels of radioactive deposition were made in most countries, but there are immense problems in trying to measure deposition across the whole of a country, let alone all of Europe. Most countries measured radiation doses, beginning early May. But it was difficult to get a consistent picture of the level and pattern of contamination, for various reasons. Measurements were scattered over a massive area, and different things were being measured in different areas (for example, some countries measured iodine-131 levels in the air, others measured them on the ground). Moreover different measuring techniques, instruments and protocols (eg timing and frequency) were adopted in different countries, making strict comparisons impossible.

IAEA and the World Health Organization (WHO) collated measurements as they became available (see, for example, Figure 8.2), in attempts to create as coherent a picture as possible of the spread of fallout across Europe.

Although actual *measurements* of fallout patterns were thin on the ground, governments were keen to find out as much as they could (and as quickly as they could) about what the synoptic picture looked like. This information was needed in drawing up contingency plans for coping with

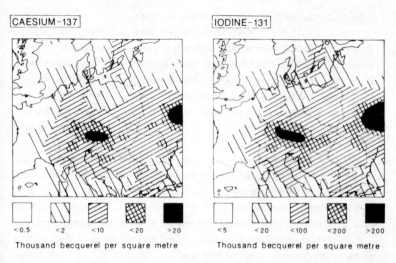

*Figure 5.4   Radiation deposition across Europe, based on model
calculations up to 6 May 1986. After Anon (1986) IAEA Bulletin (Autumn)*

the prospect of contaminated land, livestock, water and food resources,
and for dealing with the import of fresh foodstuffs and livestock from
other countries which might be contaminated. It would also be valuable
in estimating how the fallout from Chernobyl might affect people's
health in both short and long terms. The answer was provided by teams
of atmospheric scientists.

Estimates

Scientists working on air pollution problems (like acid rain) before
Chernobyl had developed various mathematical models of the atmosphere (to simulate air mixing and transport). Several groups quickly fed
into their atmospheric transport models the appropriate parameters for a
Chernobyl simulation, and produced *'best estimates'* of the levels and
distribution of deposition after the event.

The most reliable simulations were carried out by teams from Imperial
College in London, and from the Lawrence Livermore National Laboratory in California. The WHO also published a map, derived from a
simulation model, showing results for estimated deposition of caesium-
137 and iodine-131 up to 6 May 1986 (Figure 5.4). Each simulation
produces different results, as might be expected given the enormity of the
task, the shortage of good quality data to feed in, the need to produce

results fast, and the pioneering nature of the work.

Inevitably rates of deposition will be very high in the immediate vicinity of the accident, so that high rates are estimated (Figure 5.4) for eastern Russia. This obvious *'hot-spot'* spills over into adjacent eastern European countries, most noticeably across most of eastern Poland, northern Romania and (to a lesser extent) most of Bulgaria. The ban which most western European countries placed on imports of fresh food-stuffs from eastern Europe during early May (Chapter Four) was doubt-less well-founded.

Estimated rates of deposition over Scandinavia are relatively low, probably because the cloud passed overhead relatively quickly and there was little rain to wash the radioactive fission products back to earth. Most of the radiation *'hot-spots'* predicted by the model (Figure 5.4) are in central Europe - particularly in southern West Germany, western Czechoslovakia, southern and eastern Switzerland, most of Austria and Hungary, northern Yugoslavia and northern Italy.

Limitations

The mathematical models provide a useful guide to likely patterns of deposition of the radioactive fallout from Chernobyl, but their limitations must not be overlooked. Like all models, they are only as good as the data fed into them. Equally, they are only as good as the understanding of atmospheric dynamics and circulation patterns which provides their foundation. For example, before Chernobyl scientists had very little direct experience of how active radioactive materials diffused through the atmosphere over a period of days. Similarly, rapid changes in wind patterns (speed and direction) over Europe in early May made modelling of the Chernobyl fallout extremely difficult.

There is a temptation to forget that the figures and patterns are model estimates, and assume that they are measurements. The apparent preci-sion of the figures (eg shown in Figure 5.4) can encourage such deception. Reality is never so simple as even the most sophisticated mathematical models suggest. For instance, none of the prediction models is anywhere near sensitive enough to reveal marked variations in deposition rates over even short distances, of the sort picked up by Polish field measurements throughout early May.

The estimates of fallout deposition require good information on at-mospheric conditions over the period in question, and this was quite readily available from routine meteorological observations. But the modelling procedures also require good information on the radiation released at source - how much, when, radionuclides present and in what quantities? This was much more difficult to get hold of.

The Lawrence Livermore National Laboratory calculations assumed that around half of the iodine-131 and caesium-137 had escaped from the damaged reactor (40% in the initial explosion, and a further 10% over the following days). This would represent a total release of between 1.1 and 1.48 million million million becquerels of iodine-131 and around 0.11 million million million becquerels of caesium-137. These estimates are quite reasonable in the light of the official Russian estimate - given in the Soviet Union's report to IAEA in August 1986 (see Chapter Seven) - that the accident released a total of around 1.85 million million million becquerels of long-lived isotopes (a quarter within the first 12 hours and the rest over the next ten days).

INTERNATIONAL DIPLOMATIC FALLOUT

International condemnation of the way in which the Soviet Union was handling the disaster grew through the first week in May. Of prime concern was the lack of information being released on what had happened at Chernobyl, what quantities and types of radioactive material had been released, and what the risks might be to people in countries downwind. There was also growing anxiety over the lack of advice which the Russians appeared to be giving to their own people, especially in the Ukraine and Byelorussia. The apparent lack of precautionary measures being taken at other Soviet nuclear sites - particularly the other RBMK reactors like Chernobyl - was not overlooked in the west, either.

US condemnation

The United States, far enough away to preclude any real prospect of direct contamination by radiation from Chernobyl, none the less took the matter seriously. President Reagan set up a special Inter-agency Task Force, to be headed by Mr Lee Thomas (chief of the Environmental Protection Agency), to gather what information it could on the disaster. He also condemned the Soviet authority's refusal to provide a full and speedy account of the accident.

The President told the American people in his weekly radio message on 2 May that the Soviet unwillingness to be open about Chernobyl "manifests a disregard of the legitimate concerns of people everywhere. A nuclear accident that results in contaminating a number of countries with radioactive material is not simply an internal matter ... the Soviets

owe the world a full explanation: a full account of what happened at
Chernobyl and what is happening now is the least the world community
has a right to expect".

That same day Mr Vitaly Churkin, a second Secretary at the Soviet
Embassy in Washington, testified on the accident before the US Congress. His polished performance - displaying a warmly welcomed blend
of diplomatic protocol, colloquial English and Soviet humour - was well
received on Capitol Hill and widely covered by the media. But the fact
remained that he revealed nothing substantive about Chernobyl which
was not already known from official statements released in Moscow.

The United States, Britain, West Germany and Sweden joined forces
in calling for a thorough international investigation (through IAEA) into
the disaster. In the words of Mr Kenneth Baker, Britain's Secretary of
State for Environment (speaking in Parliament on 1 May), "the international community would learn valuable lessons which would benefit us
all ... I hope that Russia will understand and accept that the price to be paid
for membership of the international community, and for the trust of that
community, is an openness and a readiness to share information ... unhap-
pily this is a price which Russia has so far shown herself unwilling to
pay".

Tokyo summit meeting

The Soviet Union weathered this initial diplomatic storm and conceded
nothing. However the issue was not simply to evaporate from the public
gaze, because the radiation cloud was - quite literally - to follow
international leaders around the world.

On 4 May leaders of the seven most powerful democracies - heads of
the governments of the United States, Canada, Britain, West Germany,
France and Italy, and the President of the European Commission -
gathered in Tokyo for a summit meeting. This had been planned much
earlier and aimed to discuss measures to combat terrorism and to explore
the possibility of a bi-lateral summit meeting between leaders of the
Soviet Union and the United States. Chernobyl, so timely in its occurrence, was hurriedly added to the agenda.

The world leaders would have been quite happy to discuss the Chernobyl situation from a distance, but the problem was to surface right over
their heads in Tokyo. The Japanese Science and Technology Agency discovered - by remarkable coincidence of timing, on 4 May - that radiation
levels in the rain which was falling gently over the city were above
normal. The Agency quickly pointed out that no-one (including the world

Table 5.2 Communique on Chernobyl issued by world leaders at the Tokyo summit meeting (5 May 1986)

"1. We, the heads of state or government of seven major industrial nations and the representatives of the European Community, have discussed the implications of the accident at the Chernobyl nuclear power plant. We express our deep sympathy for those affected. We remain ready to extend assistance, in particular medical and technical, as and when required.

2. Nuclear power is, and properly managed will continue to be, an increasingly widely used source of energy. For each country the maintenance of safety and security is an international responsibility, and each country engaged in nuclear power generation bears full responsibility for the safety of the design, manufacture, operation and maintenance of its installations. Each of our countries meets exacting standards. Each country, furthermore, is responsible for prompt provision of detailed and complete information on nuclear emergencies and accidents, in particular those with potential trans-boundary consequences. Each of our countries accepts that responsibility, and we urge the Government of the Soviet Union, which did not do so in the case of Chernobyl, to provide urgently such information as our and other countries have requested.

3. We note with satisfaction the Soviet Union's willingness to undertake discussions this week with the Director General of the International Atomic Energy Agency (IAEA). We expect that these discussions will lead to the Soviet Union's participation in the desired post-accident analysis.

4. We welcome and encourage the work of the IAEA in seeking to improve international co-operation on the safety of nuclear installations, the handling of nuclear accidents and their consequences, and the provision of mutual emergency assistance. Moving forward from the relevant IAEA guidelines, we urge the early elaboration of an international convention committing the parties to report and exchange information in the event of nuclear emergencies or accidents. This should be done with the least possible delay."

leaders) would be harmed by walking in the rain, but they advised caution to those who might drink it (they should first filter it through charcoal). Low levels of caesium-137 and iodine-131 were also found in samples of vegetables, and people were advised to wash fresh vegetables thoroughly before cooking or eating them. Sales of powdered milk in Tokyo rocketed.

Discussions about Chernobyl during the summit focussed on the need for greater international co-operation in dealing with nuclear safety. They accepted the need to oblige all nations to recognize that - to paraphrase John Donne - no nuclear plant is an island. A communique was issued

(Table 5.2) on the need for world-wide agreement on an early warning system between governments on the type of accident and leakage which had occurred at Chernobyl the weekend before.

World Health Organization report

Soon after the accident at Chernobyl, many of the 33 countries in its European region appealed to WHO for advice on how best to deal with radioactive fallout. The Organization called a special meeting of radiation experts from 11 countries (east and west) to examine what the implications of the accident might be for public health. The meeting was held in Stockholm at the European Headquarters of WHO, and its report was published on 7 May.

The WHO report echoed many governments around the world in its expression of strong dissatisfaction with the Soviet Union for failing to give adequate and timely information about the accident. It called for an improved international system for monitoring nuclear fallout and for better co-ordination between nations.

Its conclusions about health risks were reassuring. There was relatively little danger from radiation outside the immediate area surrounding the power station, it insisted. Moreover, by 7 May restrictions on food imports (see Chapter Four) could not generally be justified on health grounds. Similarly other precautions outside the Soviet Union - such as staying indoors, avoiding gardening and farming, avoiding drinking rain water and taking iodine tablets - were no longer necessary. Milk generally posed no general health risk for infants, although radioactive rainfall might justify restrictions in some isolated areas.

JEUX SANS FRONTIERS

The truly international dimension of the accident at Chernobyl was further underlined within a fortnight of the accident. On 10 May fission products were detected in air samples at Oak Ridge, Tennessee. The next day Canadian government officials reported that they had detected signs of the fallout over Canada. Slight traces of iodine-131, caesium-134 and caesium-137 had been measured in air samples across the country. Rain falling over Ottawa had up to six times the permissible level of iodine-131 in it. Canadians were advised that radiation levels posed no threat to human health, but (as a precautionary measure) they should not drink rain-water until further notice.

The principle of *'good neighbourliness'* is widely established in international law. This makes it the responsibility of each country to act responsibly in the event of a major pollution problem which threatens to transcend national frontiers, and to provide its neighbours with full and relevant information, without delay. By mid-May governments around the world had grounds for accusing the Soviet Union of blatantly contravening the principle in its handling of Chernobyl. Fallout was being measured across Europe and even more further afield, and yet still the Soviet authorities said little and did even less about its invisible export of radioactivity. The shadow of Chernobyl was by now long and dark!

6
REACTIONS IN BRITAIN

The British public is divided over the issue of nuclear power. Many (including Margaret Thatcher's Conservative Government) are pro-nuclear, and happy to see expansion of the country's nuclear power programme. But there is also a strong anti-nuclear lobby (which includes the main opposition parties) which is firmly convinced that existing nuclear power stations in Britain should be phased out and no new ones built. Nuclear issues are topical, and Chernobyl happened at a time of heightened public interest and awareness. The Public Inquiry over whether to build a Pressurized Water Reactor (PWR) at Sizewell in Suffolk - the country's biggest, longest and most-expensive Public Inquiry yet - was then in full flight (see Chapter Ten).

Nuclear stories have attracted considerable interest in the British press for many years. People have grown accustomed to reading about the controversial Windscale reprocessing plant in Cumbria and the Sizewell Inquiry in particular. But the television news headlines on the evening of Monday 28 April were different - this time the talk was not of the *possibility* of a nuclear accident. One appeared to have happened! Before then the British public had never heard of Chernobyl. Since then the name itself has been used inter-changeably with 'nuclear accident'.

In this chapter we look at how Britain reacted to Chernobyl. The radiation cloud reached Britain at the end of the first week, and it found a country partially unprepared for coping with large-scale public anxiety over the invisible enemy.

RADIATION AND REACTIONS

Handling of civil nuclear emergencies in Britain is co-ordinated by the National Radiological Protection Board (NRPB) at Didcot near Oxford. NRPB were alerted by Sweden's National Institute for Radiation Protection on the Monday afternoon that background radiation levels there had risen to between two and three times the norm. The initial information gave no cause for concern, because levels over Sweden were relatively low and the cloud was not moving towards Britain.

Within hours, however, the diagnosis was to change. Swedish measurements were indicating that the accident had been much bigger than was first thought. The threat to Britain was regarded at the time as minimal, but there was concern over the safety of British nationals in the western Soviet Union (particularly the students in Kiev and Minsk).

The Board activated its contingency plans without delay. The Government would need to be kept fully informed, and appropriate decisions could not be made without reliable and up-to-date information on radiation levels across the country. Routine environmental monitoring was immediately stepped-up and NRPB began to collate the results. Monitoring was carried out by various agencies, including the NRPB themselves (with laboratories in Oxford and Glasgow) and the Central Electricity Generating Board (all nuclear power stations in Britain are equipped with radiation monitoring equipment). Radiation monitoring staff and equipment at the Ministry of Agriculture, Fisheries and Food (MAFF) were also mobilized.

NRPB were quick to seize the initiative in dampening down British anxiety over the fallout. Its declaration that Britain was safe and there was nothing to worry about was carried on all the Monday evening television and radio news bulletins, and it appeared on the front page of the leading papers on Tuesday. The Board stressed that no traces of radioactive fallout from the accident had yet been picked up at strategic monitoring stations in Glasgow, London, Belfast, Chilton (Oxfordshire), Bridgend (Mid-Glamorgan) and Shrivenham (Gloucestershire). The cloud would only arrive over Britain if the winds changed to north-easterly (blowing down from Scandinavia) and if no rain fell to wash the material down to the ground before it arrived. Even then, it would not arrive before Wednesday, at the earliest. Moreover, even if radiation levels as high as those being measured over Sweden were to appear in Britain - which was highly unlikely - there was still no cause for concern in Britain.

On Thursday (1 May) NRPB went on partial emergency status and activated standing arrangements for coping with civil nuclear emergencies. The Department of Environment (DoE) confirmed NRPB as the body responsible for co-ordinating and interpreting the radiation monitoring data around the country, and the media quite naturally turned to NRPB for information.

The big questions remained - would the radiation cloud pass over Britain, and if so when? Early in the week it had become clear that the radiation cloud was casting a dark shadow over a large and growing area

87

of Europe (Figure 5.2). Even worse, the levels of radioactivity being measured over Scandinavia, over 1,000 km downwind from Chernobyl, suggested that the accident must have been very serious indeed and vast quantities of radioactive material must have been released. It was not known at this early stage whether the leak in the damaged reactor had been sealed (it later emerged that it would not be for a further ten days (see Chapter Three)).

Political fallout

British politicians had two things on their minds on Monday 28 April - condemnation of the Soviet Union for its handling of the whole affair, and concern that the fallout might land over Britain.

The next day (Tuesday 29 April) the Soviet authorities admitted, after repeated Swedish diplomatic questioning, that an accident *had* occurred at the Chernobyl nuclear power station (see Chapter One). The matter was discussed in London in a debate in Parliament that same evening. The House of Commons was told that preliminary measurements had failed to detect any increase in radiation levels across the United Kingdom. But the Government recognized the seriousness of the accident, and agreed to provide any help which the Russians might require (this offer was turned down by the Soviet authorities (see Chapter Three)).

Members of Parliament of all parties were uncharacteristically united in their criticism of the Soviet Union for failing to notify neighbouring countries of the accident without delay. They accused the Soviet Union of wilfully neglecting established international protocol if not ignoring their moral obligations to inform their neighbours. British feelings were added to the growing European sense of disturbance over the way in which knowledge of the accident had emerged - not from the Soviet Union itself but by monitoring in other countries. This raised immediate concern over the Soviet leadership's sense of judgement and level of responsibility.

Prime Minister Margaret Thatcher was particularly sensitive about issues of nuclear safety, with the Sizewell Inquiry then under way in Suffolk. She stressed that Britain's nuclear industry has an "absolutely superb" safety record - "We have a very high standard of safety and design, construction, operation and maintenance of nuclear plant in the UK". She was to repeat this claim many times over the following months, as her political opponents were eager to seize the opportunity Chernobyl gave to dissuade the Government from pressing ahead with the controversial PWR development (see Chapter Ten).

During the first half of that week, many efforts were made to find out what had happened at Chernobyl. The questions were serious. How much and what had leaked out? How dangerous might the fallout be? What were the prospects of it blowing as far as Britain? The sinister Soviet shadow of silence persisted (see Chapter Three), and this fuelled British frustration and anxiety.

Parliament urged the Soviet authorities - through the Foreign Office in London and through the British Embassy in Moscow - to release the fullest possible information immediately.

Diplomatic exchanges

Discussions about Chernobyl continued in the House of Commons. Mr Timothy Eggar, Under-Secretary of State at the Foreign Office, was asked on Wednesday (30 April) if the radiation cloud was expected to move towards Britain. His reply was unilluminating and somewhat inconclusive - there was no danger at present, levels of radioactivity were being closely monitored and no increase had yet been detected. It did little to satisfy opposition MPs, and offered little comfort to the British public.

All eyes were on the weather - especially the movements of fronts and air masses over continental Europe. Other countries' loss would be our gain, regrettably! Intensive monitoring was continuing around the country and around the clock.

Early on the Wednesday Mr Guerman Guentsadze from the Soviet Embassy in London called at the Foreign Office to give an account of the accident. He spoke only in general terms, and provided no answer to the detailed questions which the British Government had asked of Moscow. That same afternoon Mrs Thatcher summoned the new Soviet Ambassador, Mr Leonid Zamyatin, to Downing Street for urgent talks. The Government's impatience was underlined by the unprecedented haste of this invitation, because the Ambassador had only arrived in London the day before. The Prime Minister told the Ambassador that Britain was willing to help in any way that it could, and urged him to ensure that Moscow would allow repatriation of the British students and teachers in Kiev and Minsk without delay (see Chapter Four). She also complained strongly that the Soviet Union had failed to provide other countries with the sort of technical information they needed to draw up realistic contingency plans.

That evening in Moscow Sir Bryan Cartledge (British Ambassador) was summoned to a meeting at the Foreign Ministry with the Head of the State Atomic Energy Committee, the Minister for Higher Education and a Deputy Foreign Minister. They told him that the fire in the reactor had been put out (this was incorrect - see Chapter Three). He was also advised that Britons in the Soviet Union, including the students in Kiev and Minsk, were in no danger from Chernobyl fallout.

The British Government also flew out radiation monitoring equipment to the British Embassy in Moscow, to allow staff there to collect independent measurements of levels of radioactivity in the capital (the Soviet authorities were releasing no information at all at this time). This was to prove an invaluable source of reliable information in the early days after the accident.

Also on the Wednesday the British Government set up an expert working group. It had two tasks - to monitor the situation on an hourly basis, and to advise Ministers on appropriate action. In retrospect it is clear that this 'emergency group' was toothless without a single named Minister being designated responsible for all matters to do with Chernobyl. In the early days of coping with the accident, different authorities were releasing different information and there was no-one empowered to make key decisions over what information should be made available to the press and the general public.

RADIATION CLOUD OVER BRITAIN

British anxiety had mounted towards the end of the week, as low level fallout had been measured over France and northern Netherlands. All the signs were of a progressive drift of radiation across Europe. Given the relentless march of the shadow, it was only a matter of time before the invasion of Britain, 2,000 km away from Chernobyl (Figure 1.1), would begin.

The Meteorological Office has since reconstructed the pathway which the cloud followed after leaving Chernobyl on Saturday 26 April (Figure 6.1). This shows the cloud sweeping north towards Scandinavia, moving slowly in the lower atmosphere, over the first weekend. It then turned quickly south-westwards on the Monday (steered by a high pressure zone over northern Europe and a depression over the Adriatic), and passed over Poland, Czechoslovakia, Austria and the Alps. The cloud passed over central Europe through the week (Figures 5.1 and 6.1). As the depression moved towards the north-west on Thursday (1 May), it forced the radiation cloud over France and towards the United Kingdom.

The British students evacuated from Kiev and Minsk (see Chapter Four) flew into Heathrow late on the Thursday evening, unaware that the

Figure 6.1   Path of the radiation cloud across Europe, 26 April to 8 May 1986. Based on a reconstruction using meteorological data, after Smith & Clark (1986)

radiation cloud would follow them a few hours later. As soon as they landed measurements were made of iodine-131 levels in their thyroids. These (1,900-6,500 becquerels) gave committed dose equivalents of between 4 and 14 millisieverts per person - well within the danger zone (Figure 2.3).

D-day; Friday 2 May

The south-easterly winds blew the radiation cloud from France over the English Channel, to arrive over south-east England early on the Friday

91

morning (2 May). By now it had been travelling for six days, so some of the radioisotopes with a short half-life (Table 2.3), including iodine-131, were starting to fade away. It had also crossed most of Europe, and *en route* some of the most dangerous radioactive material had already been dry-deposited or washed out by rain (for example over Poland on the Wednesday (see Chapter Four)).

There was no doubt that the cloud had arrived because before mid-day it had been detected on radiation monitors across south-east England. NRPB picked it up on their equipment at Chilton (Oxford), and the Ministry of Defence monitored it at Aldermaston (in Berkshire). It was also detected at CEGB power stations at Dungeness (Kent) and Winfrith (Dorset) and measured by the Atomic Energy Research Establishment (at Harwell, near Chilton). The press were quick to print the story of how the radiation had also been detected on makeshift equipment at Ackworth School in Yorkshire, some time before official readings were released.

The cloud's movement over Britain was closely followed by the Meteorological Office (Figure 6.2). It crossed the Kent-Sussex coast at about 5.00 am (GMT) on Friday morning, continuing to drift further

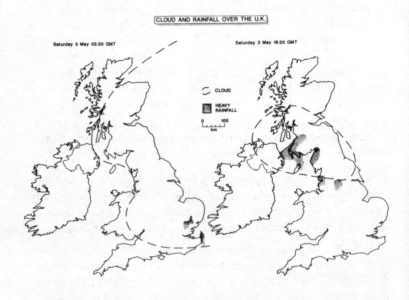

Figure 6.2   The Chernobyl cloud and rainfall over the United Kingdom, 2-3 May 1986. After Smith & Clark (1986)

northwards through the day. By evening it had reached the north of England. The only areas not covered were Cornwall, Devon, west Wales and north-west Scotland.

Friday was dry and sunny, but conditions changed on the Saturday as the low-pressure cell to the south-west deepened. This triggered off a series of thunderstorms, first in the south-east (early in the day) then over East Anglia and the North Sea. New storms developed over central England, and these moved slowly northwards to Cumbria and south-west Scotland. These storms caused the radiation cloud to shrink (Figure 6.2) as the zone of convergence effectively 'sucked in' surrounding air.

By early Sunday morning the cloud had left most of Britain, and by 9.00 am (GMT) it had passed over the Outer Hebrides. It blew out to sea over the eastern Atlantic for three days, then returned over south-west England, much diluted, on the Wednesday (7 May). By Thursday it had floated over central and north-east England, *en route* to western Scandinavia. By now the 'cloud' was very diluted and dispersed, so radiation levels within it were much lower.

Fallout

Most of the country escaped any measurable degree of deposition as the cloud drifted ominously overhead. Only where radionuclides were washed out by rain were there problems. Measurements taken between 2-5 May showed huge differences in levels of radioactivity between dry air and rainwater - air had between 1-10 becquerels per cubic metre of iodine-131 and 0.5-5 becquerels per cubic metre of caesium-137; rainwater samples had between 10-10,000 becquerels a litre of iodine-131 and between 20-2,000 becquerels a litre of caesium-137.

The most intense storms on the Saturday afternoon were centred over Cumbria in north-west England and over southern Scotland (Figure 6.2). Both areas have extensive hill-farming, and cattle and sheep were out in the fields and on the fells grazing when the fallout came.

Once radioisotopes are deposited on the ground, they can get taken into vegetation (such as grass) and become locked into soils for long periods. This creates the potential for widespread human exposure to radiation, via ingestion of contaminated foodstuffs (see Figure 2.2). Sheep and cattle eat contaminated grass and herbs, and the radioisotopes accumulate within their bodies. Meat is thus contaminated. Vegetables growing out-of-doors while the cloud was overhead can also be contaminated with dangerously high levels of radioisotopes.

It soon became clear that the ban on food imports from countries around the Ukraine (see Chapter Four) might have to be complemented by restrictions on the movement, sale and consumption of home-grown food from some areas within Britain. MAFF were responsible for monitoring levels of radioactivity in British meat and vegetables. Port Health Authorities throughout the United Kingdom were instructed by the Department of Health and Social Security (DHSS) to set up surveillance stations to monitor all food shipments from the Soviet Union, eastern Europe and Scandinavia (see Chapter Four). Particular attention was paid to fresh Polish fruit and vegetables, which are normally imported in large quantities for processing in Britain. Any load with detectable radiation levels was turned back.

Contaminated milk, foodstuffs and water

Exposure to radiation is also increased by drinking contaminated milk. Cows graze contaminated grass, and their mammary glands effectively concentrate the dietary iodine into their milk. The problem is two-fold - milk provides a quick route for iodine to be passed through pastures and dairy cows to humans, and most children (a high risk group susceptible to radiation sickness in any case) drink milk in abundance. In fact most human ingestion of iodine-131 comes from drinking milk.

The authorities fully appreciated the need to closely monitor levels of radiation in fresh milk around the country. MAFF was responsible for this too. Milk samples from different areas were collected daily after Thursday 1 May, and tested at MAFF's Central Veterinary Laboratory at Weybridge in Surrey. Milk from the high rainfall areas (Figure 6.2) was examined particularly closely and regularly.

The first signs of increased radiation levels in British milk came on the Sunday (4 May). A small but measurable rise in levels of iodine-131 in some milk samples was reported, along with traces of some other isotopes of iodine. MAFF were quick to put the public at ease, stressing that all of the measurements were well within safety limits and no restrictions on milk sales were likely.

MAFF gave further details of milk contamination the following day (Monday 5 May). These confirmed that levels of iodine-131 in British milk samples had risen over the previous three days, being generally in the region of 20 becquerels a litre. But these were still well within safety limits - the so-called *Derived Emergency Reference Levels* (DERLs) - set by the International Commission on Radiological Protection (ICRP). All necessary precautions were being taken to ensure that no contaminated milk would be distributed to consumers and milk testing would continue.

People in Britain were anxious to find out how safe milk was, and the issue of milk contamination was widely discussed in the press. It soon emerged that Britain differs slightly from most other countries in its definition of *'safe'*. The internationally accepted *'emergency levels'* for milk (given by ICRP and accepted by IAEA) are 10,000 becquerels a litre for adults, and 1,000 becquerels a litre for infants. In Britain the figures used by NRPB are the same for adults, but 2,000 becquerels a litre for infants. No official reason for the discrepancy in infant levels was given, and this odd inconsistency puzzled and worried a great many people.

Detailed monitoring of activity levels in Cumbria was carried out by the Institute of Terrestrial Ecology (ITE), a government research station at Grange-over-Sands. ITE data show that by mid-May, nearly one in five samples of Cumbrian grass had extremely high levels of a range of radionuclides. One sample of grass had readings of 44,000 becquerels a square metre for caesium-137 and a further 24,000 becquerels a square metre for caesium-134. The NRPB recommends that cattle should be removed from pasture if activity exceeds 13,000 becquerels a square metre, yet Cumbrian farmers were not advised to take their cattle in by any responsible body (like NRPB or MAFF).

MAFF also announced on 5 May that foodstuff testing would be extended to include vegetables. The timing of the Chernobyl accident had been fortunate in this respect, because little home-grown produce was exposed to fallout. Most winter greens (like cabbages and broccoli) had already been harvested, and it was too early for summer vegetables (like peas, beans and outdoor lettuces). MAFF laboratory facilities were also made available for testing levels of radioactivity in foodstuffs imported from eastern Europe.

The notion of food testing was warmly welcomed, but the details of how this was done caused some concern. It soon emerged that national decisions concerning the safety of British vegetables were being made on the basis of a mere 50 samples taken from around the country. MAFF stepped up the sample size to 100, but this still seemed an inadequately limited foundation of fact on which to base such serious decisions.

As reports circulated around Europe that drinking water supplies in the Ukraine were in danger of wholesale contamination from Chernobyl fallout (Chapter Three), people's thoughts started to turn towards the more immediate problem of British water resources.

NRPB announced (Monday 5 May) that there was little risk of water from reservoirs and taps being contaminated in Britain. The only area of slight concern was over the drinking of rainwater from parts of north-west England, Wales and Scotland where the rain had washed the radioactive material down to the ground on Saturday (Figure 6.2). No formal restrictions on drinking rainwater were imposed, but advice was given to people in Scotland, north-west England and north Wales not to drink fresh rainwater continuously for the next week. The advice was aimed mainly at people who were camping and caravanning over the holiday weekend, who might have otherwise drunk rainwater from tent canopies.

Levels of radioactivity in some areas were amazingly high. The largest readings were taken in Scotland, where levels of iodine-131 in drinking water in some areas came close to the *Derived Emergency Reference Level* (DERL) of 10,000 becquerels a litre. Physicists recorded readings of 26,500 and 28,000 becquerels a litre in a roadside puddle of rainwater near the Dounreay nuclear power complex in Caithness (Scotland) on Friday 9 May - over two and a half times the recommended safety level.

Levels of activity in rainwater declined as the cloud moved away from Britain. Throughout the episode, levels in reservoirs and tap water supplies remained well below danger levels. The effect of the radiation cloud on water from mains, streams and wells was dismissed by NRPB as 'insignificant'. On 6 May Parliament was advised that levels of radiation in piped water were less than 1% of the level at which special action is required.

INFORMATION SYSTEM

By the first weekend (3-4 May) people throughout Britain were well aware that the shadow of Chernobyl was now affecting them. The initial shock that such a large nuclear accident had happened anywhere (tinged with relief that it had happened so far away) had changed to mounting anxiety over how they personally might be affected. Concern was most acute amongst parents, pregnant women and farmers who all wanted advice on what to do.

Telephone lines to the NRPB and MAFF quickly became overloaded as the public craved more information. The information system quite literally became overloaded. There was the added problem of uncertainty over just what were the implications of the radiation measurements being reported daily in the media. To all but the nuclear scientist the terms used and their precise meaning were unfamiliar and difficult to understand.

Uncertainty and confusion were added to the frustration of delay in finding out. It was a recipe for widespread public anxiety and growing frustration at the lack of guidance on how best to ensure that personal risks from the fallout were minimized. The authorities were faced with a tidal-wave of requests for information and advice, and they just could not cope.

On Monday (5 May) the Ministry of Agriculture announced the setting up of a special operations room in London to answer questions from the public and provide advice on foodstuffs. The Department of Environment acted as overall co-ordinator and provided a clearing house for relevant information. The Technical Information Centre was set up in the Incident Control Room at DoE headquarters in London. It was manned by civil servants and government scientists (from DoE, MAFF, DHSS and NRPB).

The Technical Information Centre was also heavily over-subscribed with requests for information and advice, and many people gave up trying to get through after facing interminably long delays in getting a line in that was not already engaged. It issued ten information bulletins about radiation levels and foodstuffs, and it published two sets of monitoring results (which had been compiled by NRPB).

The authorities recognized the pressing need to not only make information available as quickly as possible, but also to make it as intelligible as possible. It had to be released in a form that the general public could readily understand to minimize the risk of wholesale panic caused as much from misunderstanding as from lack of information.

Orchestrated efforts were made to play down the seriousness of the cloud and its passage over Britain. Government scientists stressed that the remnant which had reached Britain was much diluted, and there was no genuine cause for concern. NRPB advised that levels of radioactivity over Britain were much lower than those being registered over Sweden a few days earlier. Sir Donald Acheson, Government Chief Medical Officer, insisted that there was no danger to the health of people in the United Kingdom.

These passifying statements did work, to some extent. There was no public panic to buy iodine solution in Britain, and the main impacts of the whole affair on health were neuroses and anxiety rather than any detectable degree of radiation poisoning.

LONG-TERM PROBLEMS FOR FARMERS

Radiation levels in Britain were never high enough to put human health at direct risk, although a careful watch for contamination of milk, water and foodstuffs was needed throughout May 1986. The main losers from Chernobyl will be farmers whose activities were restricted as a result of the fallout, principally in the high rainfall areas of Cumbria, south-west Scotland and north Wales.

Dairy farmers have a legitimate grievance, because they were not advised by MAFF to keep their cattle indoors during the storms on 3 May and to feed them on uncontaminated fodder subsequently. Simple precautions like these would have ensured that cattle were exposed to minimal amounts of radioactive fallout, and the widespread public anxiety over milk contamination might have been easily avoided.

The accident created serious problems for Britain's upland sheep farmers. Sheep were grazing freely on the upland moorlands of Cumbria, Scotland and north Wales when the rain washed fallout down on the Saturday (3 May). They quickly accumulated high doses of caesium and iodine from eating the contaminated vegetation. The NRPB emergency reference level for meat is 10,000 becquerels per kilogram of flesh, but the British Government had adopted an *'action level'* of 1,000 becquerels a kilogram after an *ad hoc* meeting of European Community advisers in early May.

As soon as measured levels reached the *'action level'* , Government action to prevent public consumption of contaminated meat was regarded as necessary. Measurements were taken by MAFF from early May onwards, and many showed unacceptably high radiation levels. For example, caesium levels in a sample of Cumbrian lamb taken on 14 May were as high as 2,450 becquerels a kilogram - two and a half times the *'action level'*.

Yet Government Ministers took no action. One explanation for deferring the decision on meat consumption was that the authorities hoped that levels of radiation would only remain above the *'action level'* for a short spell. Panic reactions could have followed from an announcement that lamb slaughter was to be restricted because of Chernobyl, and the Government wanted to avoid panic at all costs.

But radiation levels did not drop sharply, as hoped. It eventually became clear that restrictions were inevitable. On 20 June Michael Joplin (Agriculture Minister) announced that measures were being introduced to control the movement and slaughter of sheep in south-west Cumbria and north Wales. Four days later the ban was extended to cover Scotland.

The ban angered upland sheep farmers in a number of ways. Many were suspicious of the Government's ability to monitor radiation levels and adopt suitable controls. Sheep farmers in north Wales pointed out, for example, that they had been told not to send lambs for slaughter, but they were still allowed to sell milk (for example, for cheese making) from the same sheep.

Compensation for loss of income caused by the ban was utmost in everyone's mind. The original intention was that MAFF would pay compensation to farmers who were able to show that they had suffered financial loss by having to keep lambs off the market for so long that they no longer qualified for the premium prices paid for lean animals. There were no plans to compensate farmers for reduced prices for the sale of lamb from farms outside the two affected areas.

MAFF anticipated being faced with a total bill of around £10 million for compensation. The logical next step would be to pass this cost on to the Soviet government - the accident had happened on *their* territory, after all. But the matter of reclaiming the cost was left in the hands of the Foreign Office (who had made no material progress in securing payment from the Russians two years later).

Radiation levels in grass, vegetation and soils have not declined as fast as scientists had predicted (and farmers had hoped). One reason was the waterlogged, peaty nature of the soils in these upland areas. These lack the clays found elsewhere, which would have 'bound up' the caesium (locked it chemically) into the soil and make it inaccessible to growing vegetation and grazing sheep.

Farmers have also been angered by the longevity of the ban. It was originally intended to last for 30 days, but it was still in force in Wales and Cumbria on the first anniversary of the accident. Over 150 Cumbrian farms were still affected by the ban on lamb sales by the end of April 1987, and between them they expected over £3 million in compensation payments by the end of 1987. Many of the restrictions on farmers were still in force over two and a half years after the Chernobyl incident.

The long shadow of Chernobyl is affecting British sheep farmers in another way too. High radiation levels were measured in spring lambs born in 1987, the first generation to be born since the accident. The belief is that caesium-137 is being passed on to the new lambs in their mothers' contaminated milk.

Scientists are not agreed on how long the high levels of radioactivity might persist in Cumbria and north Wales. Some believed that it would decline by the end of 1987 (which most certainly did not happen), whilst others have maintained all along that it is likely to stay around until perhaps the turn of the century. If the latter is true, then the long-term prospects for many of Britain's upland sheep farmers - amongst the most marginal sector in agriculture at the best of times - are bleak. They are the ones who stand to pay most of Britain's ultimate cost of Chernobyl.

DEPOSITION OF FALLOUT OVER BRITAIN

The detailed monitoring system which was activated on Tuesday 29 April, while the cloud was still over central Europe (Figure 6.1), collected a vast amount of information on radioactivity from the Chernobyl fallout. This enabled a fairly detailed picture to be built up of the patterns of fallout over the country. The likely effect of the fallout on people in the United Kingdom could also be assessed.

NRPB data suggest that the average person in the UK would be exposed to around 0.07 millisieverts of radiation from Chernobyl between May 1986 and April 1987. This would add just over 3% to the average exposure received from natural background sources (2.15 millisieverts, see Table 2.1). It would be impossible to detect any long-term health consequences (eg increased rates of cancer) from such small additional exposures.

But the pattern is not uniform across the country, mainly because more radioactivity was washed from the cloud when it rained in the north (over some parts of Wales, northern England and Scotland) on the Saturday (Figure 6.2). The estimated increase in annual exposure is around 15% in the north, and roughly 1% in the south.

The calculations suggest that direct exposure to the fallout in Britain will have been minimal. External irradiation is estimated in the region of 0.0001 millisievert a year for most people, an order of magnitude lower than what we normally receive as background radiation from the nuclear industry (Figure 2.1). Similarly the amount of radioactivity breathed in was extremely small. Inhalation is estimated at between 0.007 and 0.008 millisieverts over the year for the average person.

The difference between south (dry) and north (wet) has little bearing on direct exposure and inhalation. But it does affect the amount of contamination people received from direct irradiation from the ground. Contamination from beta rays is estimated at between 0.00003 millisieverts (south) and 0.001 millisieverts (north) over the year. Contamination from gamma radiation is estimated at between 0.01 millisieverts (south) and 0.16 millisieverts (north). None are significant in absolute terms.

The north-south divide is also reflected in the pattern of variations in activity levels in soils, grass and foodstuffs. Levels of radioactivity measured on grass in the north of England were much higher than those measured in the south; generally in the order of a few thousand becquerels per cubic metre (north), compared with a few hundred (south).

Concentrations of iodine-131 in cows' milk show the same trend, with values in the order of 400 becquerels a litre in the north compared with around 50 becquerels a litre in the south. Caesium levels varied similarly (Table 6.1). All reported values were well below the critical level at which NRPB advises a ban on the sale and consumption of milk.

Activity levels in fresh vegetables exposed to the fallout while growing were also much higher in the north (Table 6.1), but again were nowhere near high enough to justify limits on consumption.

Table 6.1 Peak activity levels in the UK

Radionuclide	Milk (becquerels per litre)		Vegetables (becquerels per kg)	
	south	north	south	north
iodine-131	50	400	100	200
caesium-134	4	400	10	100
caesium-137	2	200	5	50

SOURCE: Fry *et al* (1986)

Exposure to radioactivity varies quite a lot between children and adults, because they have different eating preferences and patterns of activity. It is estimated that adults take in roughly 40% of their dose through inhalation, 20% from digestion and the rest by contact with deposited (gamma ray) particles. Children tend to take in up to 60% through their diet (in which milk and milk-products figure prominently), around 14% by breathing and the rest from particles.

Consequently exposure rates can differ between adults and children, even in the same family in a given area. The NRPB calculations suggest that children in the south might have an added dose of around 0.05 millisieverts, compared with an extra dose of up to 0.9 millisieverts in the north (the difference largely reflecting activity levels in milk).

Most of the additional radiation doses to adults and children will have been received within the first month or two. Some of the radionuclides - especially iodine-131 - have such short half-lives (Table 2.3) that they soon start to decline in activity. In addition some of the deposited nuclides would quickly be washed from the ground into rivers and lakes. Through time, therefore, the exposure risk should decline progressively (but the scientists are uncertain about the speed at which this is likely to occur).

Total doses to individuals are likely to rise through time, as they come into contact with more contaminated material (even if the levels of activity in it decline through time). It is estimated that over the five years following the accident (ie up to May 1991) total radiation doses from Chernobyl are likely to be in the order of 0.4 millisieverts in the north and 0.02 millisieverts in the south. Set alongside expected doses from natural background sources of radiation (Table 2.1) these seem remarkably low, even in the north.

LONG-TERM HEALTH RISKS

Despite the public anxiety in Britain over Chernobyl and its fallout, the real risk to health appears to be very small. A number of studies have examined likely absorbed radiation doses, and they all conclude that, when viewed alongside natural background sources of radiation (see Chapter Two), the accident is likely to have minimal long-term impact in Britain.

A detailed study was carried out on the area around Berkeley in Gloucestershire, using measurements of radionuclide concentrations in the air, on the ground, and in rainwater, tapwater, grass, milk and green vegetables. The aim was to estimate the total radiation dose people are likely to receive from Chernobyl through direct and indirect pathways (see Figure 2.2). The results suggest that the average adult will have received an effective dose equivalent to around 0.2 millisieverts, around 10% of the average exposure to background radiation in the UK (see Table 2.1). A one year old child will receive around 0.5 millisieverts (about a quarter of the background level), it is estimated, and a ten year old child would receive something between the two.

National estimates reflect this local area study. It is estimated that over the first 26 days after the accident, doses of iodine-131 absorbed in the thyroid tissue would be in the order of 0.07 millisieverts for adults and 0.3-0.5 millisieverts for children. This compares with the estimated 1 millisievert a year of iodine-131 which we receive from natural background sources, and the estimated 0.3 millisieverts which adult thyroids might have absorbed as a result of the 1957 Windscale graphite fire (see Chapter One).

NRPB calculations suggest that fallout from Chernobyl is likely to increase the average radiation dose to people in the United Kingdom by around 3% in the next few years. As a comparison, the Board suggest that this would be roughly equivalent to the risk associated with smoking half a cigarette over the next year! At worst tens of cancers directly linked with Chernobyl fallout are likely to emerge in Britain over the next 50 years.

When viewed alongside the many millions of normally occurring cancers over the same period, Chernobyl victims will be impossible to detect. Moreover, the risk is small when set alongside other likely causes of premature death - like Aids, smoking and vehicle accidents.

While the risk estimates point to a long-term human cost which is minimal relative to some other possible causes of death, it is fair to ask whether tens, even spread across the country, is an acceptable cost. It is little comfort to those whose lives will be needlessly shortened by Chernobyl, or to their loved ones, that they are part of the 'minimal' British casualty list. Why should they be part of the shadow of Chernobyl?

BRITAIN'S ABILITY TO COPE

News reports continued to appear each day about fallout over Europe and activities around Chernobyl. The accident and its aftermath were headline news and prime-time television. Reports continued to appear on levels of iodine in British milk - yet no restrictions were imposed and little action other than monitoring seemed to be under way (at least until the lamb restrictions were announced on 20 June).

This lack of activity puzzled many observers, who called into question the British Government's level of preparedness for and ability to cope with a nuclear accident of this sort. NRPB announced on Monday 5 May that "there is some alarm among the public and we are doing all we can to answer all the inquiries". But that was not enough for many people.

Friends of the Earth (FoE), the anti-nuclear environmental group, insisted that the Chernobyl affair had shown the UK emergency information system up as inadequate and incapable of answering even the simplest requests for information. FoE also claimed that it highlit how impossible it would be for the public to make informed judgements on the basis of the generalized information being released by NRPB and MAFF (the two key organizations involved).

The whole experience led many people to doubt the practicability of existing Government arrangements for dealing with nuclear accidents. NRPB is ultimately responsible for all radiation monitoring in Britain, but it effectively sub-contracts part of this responsibility, to (a) the Meteorological Office, which checks radiation levels in the upper atmosphere on behalf of NRPB, (b) MAFF which monitors radiation levels in foodstuff and drink (including milk), and (c) the CEGB which collects ground level radiation measurements at nuclear power stations.

If radiation levels rise significantly, NRPB informs ministers at the

Department of Environment (DoE), the Department of Health and Social Security (DHSS) and the Ministry of Agriculture, Fisheries and Food (MAFF). It is the joint decision of the ministers on whether or not to recommend to Government that milk supplies and other foodstuffs be banned. NRPB insisted that no radiation levels sufficiently high to trigger this ministerial recommendation procedure had been detected in British milk or in air or on the ground in Britain since Chernobyl.

But by early May there was growing concern amongst parents, pregnant women and farmers in Britain. Many calls were made to the NRPB and to MAFF from people wanting to know how safe food and water were. The information system was quite simply overloaded. The 20 line switchboard at NRPB was almost permanently engaged during early May, and the Board estimate that over 10,000 incoming calls were made on Sunday 18 May (most of which could not be connected).

There was an added problem, of uncertainty about just what the implications of some of the radiation measurements really were. NRPB have since argued that facts were made readily available to people. What they really wanted when they phoned in was some personal contact with someone who would listen and give personal assurances that all was well. The Board found itself in the unusual position of acting as a nuclear counselling service.

Uncertainty and confusion were added to the frustration of delays in finding out. It was altogether a recipe for widespread public anxiety and growing frustration at the lack of guidance on how they should best react to ensure maximum safety for themselves and their families. It has been argued that in Britain, fear of radiation did much more harm than radiation did itself. There are salutory lessons for contingency planning and information provision in times of widespread public anxiety (see Chapter Eight).

GOVERNMENT REPORT ON CHERNOBYL

Some of the apparent complacency with which the British Government handled criticisms after Chernobyl that foodstuffs may have been contaminated by radiation fallout was to disappear after the publication in late July 1988 of the House of Commons Agriculture Committee report *"Chernobyl; The Government's Reaction"*.

The report stated that there was no convincing evidence that public health had been put at risk, although it was likely that some lamb with radiation levels above the statutory limit had been eaten by the public after Chernobyl. Contaminated lamb could have reached the shops by several routes, and the report concluded that "it must therefore be probable that some did so". However Mr Jerry Wiggin, Conservative chairman of the committee, stressed that "someone would have had to have an exclusive diet of contaminated meat for a lengthy period for any harm to come about".

The imposition of first restrictions on slaughtering came seven weeks after the Chernobyl fallout landed over Britain, and some young suckling lambs (the group which showed the highest radiocaesium levels in MAFF monitoring) could have been slaughtered over that period. Moreover, the Scottish Department of Agriculture tried to minimize disruption to Scottish farmers and in doing so imposed stock restrictions which were not entirely watertight. In addition, restrictions were not placed on some marginal areas until 1987, and it was quite likely that some sheep from these areas had been contaminated but entered the foodchain.

The report also says that children should have been advised not to drink milk immediately after the fallout, a precaution taken in several European countries (see Chapter Four). Countries that did take such precautions were able to reduce radiation exposures of 'high risk' groups to levels lower than in the UK.

Another worrying revelation in the report was the existence of a *'hot spot'* of high contamination on the ground near Skipton in North Yorkshire which had not been detected earlier (and which consequently does not appear on Figure 6.2). It appeared to be caused by a local heavy rainfall on the Saturday (3 May 1986), and could well have contaminated local lamb and people. The Meteorological Office had apparently warned the Ministry of Agriculture in June 1986 that there had been heavy rain in that area while the radiation cloud was overhead, but the Ministry failed to act until November 1987 when it found two sheep above the radiation limit. The committee argued that "there was a clear case of negligence" in the Government's failure to impose restrictions in this area.

The committee concluded that the Government has been on the whole successful in its objective of ensuring safe food in Britain, and that it had "got it right for nearly everybody". Some 9,000 farmers had already been compensated by July 1988, but they called on ministers to find ways of making adequate payments to the 130 or so farmers whose apparently legitimate claims were then still outstanding and who were the victims of 'rough justice'.

The report is also critical of the Government's response to the Chernobyl emergency, and recommended changes in future procedures. It proposed, for instance, that contingency plans to protect public health after any future nuclear accident to avoid the sort of confusion and delays in supplying information which added to public anxiety over Chernobyl. The Ministry of Agriculture had been responsible for food supplies, the Department of Environment for water and the Department of Energy was assumed to have overall knowledge of nuclear contamination. The report also recommended that in future local authorities should become involved in national contingency plans, and be encouraged to develop their own monitoring facilities.

The matter of public relations, awareness and information was also addressed, and the committee stressed the need for Government to pay attention to the need to allay public fears about radiation fallout. This, they concluded, "can raise the level of debate to everyone's advantage, and misconceived or exaggerated fears can be laid to rest". The problems of coping with vast numbers of public enquiries "should be addressed at a deep level and not simply in terms of numbers of telephone lines". Uncertainty over the meaning of radiation safety levels and what they mean in practice had also emerged as a real problem, and the committee argued that the Government should publicly justify its decisions about acceptable levels of radiation "in a way which does justice to the complexity of the factors which need to be taken into consideration".

In final analysis, the committee concluded the British Government had learnt the right lessons from Chernobyl for the future. It remains to be seen when, how and to what extent the Government will act on the recommendations of the report.

7

WHAT REALLY HAPPENED - THE RUSSIAN REPORT

Much of the anxiety around the world in the wake of Chernobyl arose from a lack of information about what had happened and what was happening at the stricken nuclear power station. The communique issued after the Tokyo summit meeting in early May (Table 5.2) captured world feeling - the Soviet Union should provide other countries with speedy explanations and permit international inspection of their other nuclear power stations, so that everyone might be assured that a similar accident was unlikely to happen again (in the USSR or elsewhere).

Many western observers felt that Moscow was handling the incident in customary style, with unnecessarily tight secrecy and extreme reluctance to provide even the sketchiest details of what was going on. Nuclear specialists were heartened when the three IAEA experts were allowed to visit Chernobyl (Chapter Three) and were given the opportunity of discussing the accident and the clean-up operations with leading Russian scientists and engineers.

In this chapter we shall look at how the Soviet Union handled its internal inquiries into the accident, and at what the official report had to say about what caused it. The Government Commission set up on the Monday (28 April) that news broke about the accident (Chapter One) had two tasks - to determine what caused it, and to "eliminate the consequences of the disaster". Its report was eagerly expected towards the end of May, but it eventually appeared late in July, having gone through at least four committee chairmen (all Deputy Premiers of the USSR).

The delay was itself further cause for concern in the west. For many weeks the Soviet Union drew a largely impenetrable veil of secrecy over their inquiries. Would the report, when it finally appeared, be a white-wash? Would it reveal exactly what went wrong, and apportion blame realistically? Once again, the world would have to wait and see.

FIRST RECRIMINATIONS

Tantalizing glimpses were revealed from time to time, and these were sufficient to convince western observers that human errors had played their part in both causing and subsequently mishandling the accident. These isolated pieces from the complex jigsaw were released by the Soviet authorities partly to protect their flagging image and suspect credibility at home and abroad. They quickly recognized the need to be be seen - especially by the international community - to be putting their own house in order.

The first wave of recriminations came on 7 May, when Mr Gregori Rivenko (leader of the Communist Party in the Kiev area) revealed, in an interview in the communist party newspaper *Sovetskaya Rossiya*, that some workers at the plant had not acted responsibly when tackling the disaster. He was frank, and added "you cannot hide a sin: individual workers were found who, in the difficult conditions, did not display the necessary steadfastness and willingness to be at the front line".

The real backlash began five days later, on 12 May, when three men who had headed a transport combine at the plant were publicly humiliated for delaying the evacuation programme of local residents by abandoning their responsibilities when the accident happened. The three were named in the Soviet newspaper *Pravda,* in an unprecedented display of candour. It was alleged that they displayed "political immaturity, inactivity and misunderstanding of the situation". One, a Communist Party official, was expelled from the Party for failing to face his responsibilities for looking after the refugees who left the Chernobyl area. One was severely reprimanded for the same offence. The third was reprimanded for failing to give accurate and speedy information about the disaster. All three were disgraced in public.

This was the first admission in the tightly controlled Soviet press that mistakes were made in the first few hours after the accident, when panic in the face of rapidly rising radiation levels must have been present if not widespread. Western observers had keenly awaited news from the Soviet authorities in Moscow about why they had been kept in the dark about the full scale of the accident for up to 48 hours (Chapter Three).

There were clear signs that the Soviet authorities would embark on a witch-hunt, to identify and punish those who had acted improperly or incorrectly before, during and after the accident. No-one yet knew just how thorough and searching this exercise would turn out to be.

The fallout from Chernobyl had been deposited over many countries (Chapter Five), and it was clear from the outset that the Soviet authorities could not treat the accident as simply a domestic issue. There were many international implications and anxieties, and the Soviet Union would be accountable to the world at large for what had happened and how it had handled the disaster.

INTERIM SOVIET REPORT

The early signs suggested that the Soviet authorities would take the matter seriously and give the Chernobyl accident the detailed examination people in the west thought necessary. Some time before the Government Commission published its report, Aleksei Makhukhin (First Deputy Minister of Power and Electrification) had told the official Soviet news agency *Novosti* that the accident had been due to "a coincidence of several highly improbable and hence unforeseeable failures". He added that whilst neither the design nor the construction of the plant seemed to be at fault, additional measures would be introduced there and in similar nuclear power stations to ensure greater safety in the future.

The Soviet Union published its interim report on the disaster on about 19 July (two and a half months after it happened). This said little about the circumstances leading up to the accident, and gave no details about exactly what had happened. But it did reveal that those responsible at Chernobyl at the time of the accident had been engaged in experiments which were not authorized by the licensing authorities. It concluded (without giving details) that those who were affected by the disruption caused by the accident would be compensated, and those responsible for it deserved to lose their jobs.

The interim report was given careful consideration by the Politburo during a special session over the weekend 19-20 July. Further details about the unauthorized experiment emerged. It was a sad tale of incompetence and dereliction of duty - managers and scientists at the plant had not prepared themselves properly for the experiment, neither had they coordinated their plans with the appropriate authorities (which they were legally bound to do); the experiment seemed to go ahead with minimal proper supervision; no proper safety measures were adopted during the experiment.

The Politburo were told that by mid-July the Soviet Procurator's Office had started criminal proceedings against "the persons guilty of the accident at Chernobyl". The Director at Chernobyl at the time of the accident (Bryakhanov) had been dismissed early in May. Other heads were soon to roll. Four senior officials were subjected to "rigorous Party penalties" and dismissed from post. They were Evgenii Kulov (chairman of the State Committee for Safety in the Nuclear Power Industry), G. A. Shasharin (a Deputy Minister of Power and Electrification), Aleksandr Meshkov (a First Deputy Minister of Medium Engineering), and Ivan Emelyanov (designer of the Chernobyl station and then Deputy Director of a research and development institute).

In some ways these 'famous five' were to become the 'sacrificial lambs' of the whole affair, and the Soviet authorities were happy to lay most of the responsibility for the accident on their shoulders. After all, they appeared to have rooted out the culprits whose irresponsible acts had caused the disaster. Western observers have since remained alarmed at the way in which a few named individuals have received all the blame, whilst the Soviet authorities insist that there are no basic flaws in the RBMK reactor design, nor in their reactor safety systems and philosophy (especially in terms of lack of proper containment structures) (see Chapter Nine).

IAEA POST-ACCIDENT REVIEW MEETING IN VIENNA

The International Atomic Energy Agency (IAEA) had watched events surrounding Chernobyl with keen interest, and its Board of Governors held a special session in late May to decide how best to proceed. They called for a post-accident review meeting, to be attended by safety experts from around the world, to exchange information and explore the general implications of what had happened.

The meeting was held in Vienna, at IAEA headquarters, from 25-29 August 1986. It was attended by around 500 technical experts from 62 countries and 21 national and international organizations. Daily press conferences were given to the assembled international press corps, over 200 strong.

The Soviet Union sent a 28-member delegation. It was headed by Professor Valerie Legasov, a distinguished Academician of the Soviet Academy of Sciences and Deputy Director of the I. K. Kutchatov Institute of Atomic Energy.

Russian report

Exhibit A at the meeting was the 382 page technical report from the Soviet authorities (in Russian). It had been prepared by a team of experts appointed by the USSR State Committee on the Utilization of Atomic

Energy, based on the findings of the Government Commission on Chernobyl. It described the Chernobyl RBMK reactor in detail, gave an analysis of the accident and its consequences, and described measures taken by the Soviet Union to provide medical treatment, to monitor radiation and to decontaminate the area around the nuclear plant.

Professor Legasov introduced the Russian report at a press conference on 25 August. He noted that the accident "was a disaster for our citizens. It has brought with it considerable material and moral loss. It has disturbed the current economic and scientific activity of many enterprises and organizations. It has made it necessary to work in an environment with a very complicated radiation picture. And, in the world community, this accident has led to an intensification of discussion in relation to our preparedness for a further utilization of nuclear energy".

He also provided some details on how the accident had affected Russia's nuclear power industry. About half the country's RBMK reactors had by then been shut down for safety-related modifications designed to improve reactor reliability while maintaining economic viability. The authorities were hoping to return reactors one and two at Chernobyl into service by the end of 1986, but it was not yet clear when reactor three - situated beside the damaged number four reactor, and itself threatened by fire at one stage (Chapter Three) - could be put back into service.

Construction work on reactors five and six had stopped for the time being, and their design would be altered to include appropriate safety modifications. The development of nuclear power in the Soviet Union would continue, noting the lessons to be learned from Chernobyl.

Unauthorized experiment

The report gave full details of how and why the accident had happened. It did not shy away from pinpointing areas of human error or incompetence and from assigning blame.

The accident was described as "the culmination of an almost incredible series of errors" by a group of unnamed technicians. It happened, paradoxically, in the course of a (admittedly rather unorthodox) safety test. The scientists were trying to establish how long the reactor's turbine generators (see Figure 1.2) would continue to turn if the reactor had to be quickly shut down. Put simply, they were seeking to establish if the turbo-generator would continue to rotate (thus supply electricity) for 30-45 seconds after its power supply was cut off. If it did, then back-up diesel generators could be started up relatively slowly, with consequent gains in efficiency and reductions in potential damage to the equipment. If it did, the residual energy could also be used to run cooling pumps in an emergency.

The experiment could only be done while the reactor was being shut down, and the opportunity came during the scheduled overhaul of number four reactor at the end of April. This deadly 'comedy of errors' involved six critical but deliberate actions by the technicians involved (Table 7.1), who effectively disconnected the entire automatic safety system within the reactor to prevent it from interfering with the smooth running of the experiment.

There was no detailed record of how the reactor behaved during the experiment, because the data recording system used in monitoring routine operations had been switched off by the operators. The sequence of events had to be reconstructed afterwards, making much use of Soviet mathematical models of the reactor.

Table 7.1 The six critical errors that caused the explosion

ERROR ONE	emergency cooling system was turned off
ERROR TWO	reactor power output was inadvertently lowered too much (making it difficult to control)
ERROR THREE	all water circulation pumps were turned on (exceeding recommended flow rates)
ERROR FOUR	the automatic signal that shuts down the reactor if the turbines stop was blocked
ERROR FIVE	the safety devices that shut down the reactor if steam pressure or water levels become abnormal were turned off
ERROR SIX	almost all control rods were pulled out of the core

SOURCE: summarized from Serrill (1986)

Sequence of events

Test preparations began at 1.00 am (local time) on Friday 25 April, when the reactor was running at normal power (producing around 3,200 megawatts of thermal energy). The operators began to reduce the operating power level in the reactor to half its normal level, in preparation for the scheduled routine maintenance shut-down - and to make their secret experiment possible. By about 1.05 pm, power was down to around 1,600 megawatts and one generator was then switched off. The plan was to run the experiment on the second generator when the power level had fallen to between 700 and 1,000 megawatts.

The reactor has an automatic emergency water cooling system which is designed to prevent it from going 'out of control' by having an uncontrolled fission chain-reaction (see Figure 1.3). This cooling system was switched off at 2.00 pm to make it possible for them to operate the reactor at low power levels. This was a gross violation of operating instructions, and the first critical error (Table 7.1).

The test was set back 9 hours by an unexpected local need for electricity, which made it necessary to keep the reactor running. The emergency cooling system was not switched on during this period. By good fortune no mishaps occurred, although the operators knew full well that they were breaking well-established safety rules.

The operators began their test sequence at 11.00 pm. Their first task was to reduce the power level further, down to 25% of the norm. For some unspecified reason (probably incompetence; the Soviet report speaks of technically inexperienced personnel) they badly overshot the target, and the power level dropped suddenly down to around 1%. It was extremely difficult to stabilize the reactor at this stage - the control system had been disengaged and xenon gas (a by-product of iodine-131 decay, which slows down the fission chain-reaction) had built up naturally in the reactor.

The operators should have stopped the experiment at this stage, but they pressed ahead. They needed to raise the power level (to at least 7%), and they could only do this by withdrawing most of the control rods from the reactor (see Figure 1.4). There should have been 30 control rods in the reactor at all times, to regulate the fission chain-reaction; the operators removed all but 6 or 8.

By now the reactor was running with no protection, no control systems and no emergency cooling system. The operators had managed to stabilize the power level in the reactor, but they had lost control of it. It was only a matter of time before fate would strike, but the scientists were too preoccupied with the experiment to give more than passing thought to the prospect of serious failure.

By 1.23 am on Saturday 26 April, conditions were right for the test to begin. Power was stable at about 200 megawatts, and the plant supervisors decided to proceed with the experiment. They connected two more pumps to the reactor, and straight away too much cooling water flowed through the reactor core. The operators then blocked signals from pressure and water-level sensors, to prevent an automatic shut-down of the reactor (which would have terminated their experiment). This effectively disabled a key part of the reactor's emergency shut-down system.

The flow of power to turbine number eight was then stopped, and the operators held their breath - as well they might.

The reactor immediately began to overheat dangerously. In some water-cooled reactors (especially the RBMK design), unexpected boiling of water (which turns instantly into super-heated steam) can cause an increase in the nuclear chain-reaction rate. This releases more heat, causes more boiling, and so on; the reaction started quickly runs out of control. In technical terms the RBMK reactor is said to have a "large positive void coefficient of radioactivity".

The shift manager tried to introduce emergency procedures, but by now all of the reactor's defences had been deliberately disabled. The reactor was defenceless, and the operators knew that only too painfully. At 1.23:40 the shift manager gave the command to quickly reinsert the control rods they had taken out, but it was too late. The power output continued to rise, the core got hotter and hotter, fuel started to disintegrate and fall into the cooling water.

Within seconds there was a terrific power surge (up to perhaps 1.5 million megawatts, according to the mathematical estimates). Pressure levels within the reactor rose sharply, cooling channels were ruptured and two thermal (rather than nuclear) explosions followed. These ripped open the reactor core, blew the roof off the reactor building, started over 30 fires around the plant and allowed radioactive material to leak freely out into the air. The rest was by now history.

The report revealed that it took 90 minutes for fire engines to arrive from Pripyat and Chernobyl. By 5.00 pm most of the fires were out and number three reactor had been shut down. Reactors one and two were shut down early the next morning.

The early speculation over the cause of the accident (Chapter Three) was made before anyone outside Chernobyl was aware that human error was almost entirely to blame. The attraction of carrying out the unauthorized experiment had blinded the plant operators to their responsibilities - they were later accused of not preparing the experiment properly, not securing the approval of the appropriate authorities to run the experiment, not controlling it adequately when it was under way, and not taking the requisite safety measures.

Radiation levels at Chernobyl

The report gave some details of the amount and nature of the radiation released from the damaged reactor core, based on measurements made at the time. The explosion sent a plume of debris over 1,200 m into the air, which was soon blown across the western Soviet Union and Poland (Chapter Four) and subsequently across most of Europe (Chapter Five).

An estimated 1,850,000 million million becquerels (50 million curies) of radioactivity was released between 26 April and 6 May. About half of the total release came between 2 and 5 May, as decay heat in the remaining fuel in the damaged reactor increased temperatures in the core and caused radioactive fission products to be carried high into the atmosphere (on rising thermals). Radiation releases fell markedly after 6 May.

Radiation levels were inevitably exceedingly high in the power station complex itself. There was a sharp rise within the first few hours of the accident. The wind blew fallout over Pripyat, and by 7.00 am on 27 April (the day after the explosion) radiation readings there were between 1.8 and 6 millisieverts an hour. By 5.00 pm that day, levels as high as 7.2 - 10 millisieverts an hour were being recorded in Pripyat - before the town's 45,000 residents were evacuated (Chapter Four).

The report estimated that Pripyat received between 0.015 and 0.05 grays of gamma radiation and 0.1 - 0.2 grays of beta radiation. Thyroid glands received up to perhaps 0.3 grays from ingestion of iodine-131. Radiation exposures beyond the 30 km exclusion zone were much lower.

REACTIONS TO THE REPORT

The official Russian report was greeted as a frank and open description of what had happened, why it had happened, and how it was coped with. Experts in the west were impressed by its thoroughness, its spirit of self-criticism, and the promptness with which it was prepared. It was blunt and to the point. As one American scientist who reviewed the report concluded, "if it lacks some of the detail a compulsive expert might wish to find, it is far more than a compulsive cynic might expect".

It was in all senses unprecedented, and signalled a marked step forward in Soviet Leader Mikhail Gorbachev's policy of *glasnost* , or openness. It was seen by many diplomats as a significant sign of a new Soviet candour, acceptance that mistakes had been made, and willingness to have open dialogue with the west.

Some nuclear experts in the west criticized the report for failing to give enough information on some critical topics - such as the Soviet decontamination procedures. The basic design of the reactor was also criticized as old fashioned, and based on a safety philosophy which was completely unacceptable to the west (ie based on *prevention* rather than *containment*).

It was also felt that some important questions remained unanswered. Why, for example, had the technicians been allowed to carry out such a dangerous experiment on the reactor, seemingly without appreciating the risks involved? Some Soviet officials insisted that the workers involved must have thought that such an accident was impossible. It was also believed that complacency might have crept in, because the Chernobyl plant had been working well since 1983 with no sign of any problems.

The report acknowledged that the people running the test had inadequate understanding of the physics of the reactor. After reading it, the Politburo established that it was "solely the irresponsibility, negligence and undisciplined actions of the power plant staff which had led to such disastrous consequences". As we noted earlier, observers in the west were rather concerned at the apparent ease with which the Soviet authorities found it possible to dismiss the accident as the result of human incompetence.

Was the design of the reactor, and its safety systems, faulty or deficient? Professor Legasov insisted that "the sequence of human actions was so unlikely that the engineer (who designed the plant and its safety systems) did not include such a scenario in his project". In other words he did not allow for such gross violations of the design procedures. None the less the design engineer (Ivan Emelyanov) was sacked after the accident; his sin was omission rather than commission.

Were other Russian reactors at risk? Professor Legasov argued that recurrence was highly improbable, given the long catalogue of human negligence which had given rise to the Chernobyl accident. The accident was the result of six errors added together, which very effectively disabled all of the safety systems within the reactor. The authorities were convinced that it would not have happened if any one or more of the errors had not been committed. The RBMK reactor design was formally exonerated in the report (see Chapter Nine).

Would the accident lead to a change in Russian use of commercial nuclear power? The Soviet government was re-evaluating its nuclear-safety programme in the light of Chernobyl, and giving serious thought to more remote sites for new plants. The report stressed the need for nuclear power in the USSR to meet projected energy needs. There were no plans to phase out or shut down RBMK nuclear power stations in the foreseeable future (see Chapter Ten).

CONCLUSIONS

As the February issue of *Soviet Life* had noted in its article on Chernobyl (Chapter One), "even if the incredible should happen, the automatic control and safety systems would shut down the reactor in a matter of minutes". This assumed, of course, that the plant was running as planned, and all automatic control and safety systems were functioning. Accidental malfunction had been allowed for in the design of the nuclear plant by installing back-up cooling systems and other contingency facilities. No one had planned for malfunction caused by gross human error, yet here it was in black and white in the formal Russian report. It was a bleak catalogue of incompetence, deliberate risk and sheer dereliction of duty.

Whilst observers in the west appreciated the air of honesty in the public Soviet confession of human errors and mistakes, it did little to convince the anti-nuclear movement around the world that similar errors might not be made again. Neither did it do much to comfort people across Europe who had been exposed to high levels of radiation as a result of Chernobyl.

Formal Soviet investigation of Chernobyl was completed with the publication of their report to IAEA, but throughout the Ukraine and over a wider area of contaminated land, the clean up activities would continue for many months and years. The long-term consequences would affect many people over a wide area for a long time to come. Some of the more important long-term impacts are explored in Chapter Eight .

8
LONG-TERM PROBLEMS
AND LESSONS

Chernobyl was in many ways an unprecedented accident. Few events have cast such a long and dark shadow over such a wide area, created such a range of serious short-term problems, and commanded so much instant media attention and intense public interest around the world. We have explored the pressing problems of sealing the damaged reactor, evacuating local residents and providing medical attention to the injured in Chapters Three and Four. In Chapters Four to Six we saw how different countries coped with the immediate problems of fallout, exposure and contamination (of people and foodstuffs).

But there are also wider implications of the accident. The two most important are the likely long-term problems in the Soviet Union and beyond, and the broader lessons to be learned. They are the focus of this chapter.

LONG-TERM PROBLEMS IN THE SOVIET UNION

There is no doubt that the Chernobyl accident caused many problems, particularly inside the Soviet Union. But there are mixed views on exactly what these problems are, how serious they are, which are the most important and what can be done to resolve them.

One possible checklist of serious problems might be the charges brought in mid-July 1986 against the four senior officials who were described by the Politburo as "the persons guilty of the accident at Chernobyl" (see Chapter Seven). They were held responsible for a range of problems, including the death of 28 Soviet citizens, causing damage to the health of a large number of people (including the 203 cases which by then were being treated for radiation poisoning), the cost of evacuating and rehousing the evacuees and the cost of large scale prophylactic work (including medical checks on several hundred thousand people). There was also the problem of damage to over 1,000 sq km of farm-land which has since been taken out of production, and closure of farms and factories across the high fallout zone. Loss of electricity production created further problems.

It is impossible at this stage to say for certain what the complete catalogue of damage (inside and outside the Soviet Union) will look like, what the total damage bill is likely to be, and what problems stemming directly or indirectly from the accident might appear many years from now and possibly vast distances from Chernobyl. At least the following appear to be important.

Contamination

Many of the problems stem directly or indirectly from contamination with radioactive fallout from the accident. The Russian authorities have not made available any detailed observations about rates or patterns of deposition of the fallout across Soviet territory, although such information doubtless exists. So any evaluation of the contamination problem must rely on estimates and results from simulations carried out by western scientists.

Predicted patterns of fallout were discussed in Chapter Five, but some comments are appropriate here. The simulations run by the Lawrence Livermore National Laboratory in California assume an initial release during the accident of 40% of the reactor's core inventory (that is total load) within the first day. That would include about 80 million curies of iodine-131 and 6 million curies of caesium-137, and it compares with an estimated total release of 15 curies of iodine-131 at Three Mile Island (see Chapter One). A further 10% was released over the following five days, they estimate.

The evidence suggests that about half of the material released was lifted high into the atmosphere (perhaps over 6,000 m) with the initial explosion, then carried eastwards in upper air winds. The rest probably stayed low in the atmosphere and was blown initially north-westwards. Many areas reported two peaks of activity, which would fit this pattern. For example peaks were reported over Sweden on 28 April and 9 May (see Figure 5.1), and two phases (distinguished by different mixtures of radionuclides) were reported from Tennessee, USA, between 10-17 May and 18 May-13 June.

A vast amount of radioactive material was released in the accident, which must fall back to earth somewhere. The belief is that much of it fell close to the reactor site. Many nuclear scientists agree that Chernobyl probably produced as much radioactive fallout to be deposited on soil, vegetation and water around the world as all the nuclear weapons tests and bombs ever exploded. Up to half is believed to have been deposited within 30 km of the plant.

The Ukraine had been the site of one of the biggest battles in the Second World War, and its people had experienced considerable prolonged loss and hardship. But Chernobyl was a new battle for them - they must now fight the invisible, silent enemy of radiation which was likely to build up on vegetation, in soils, in milk and fresh foodstuffs, in drinking waters (see Figure 2.2).

The top priority was to seal the damaged reactor with sand, lead and boron, and then entomb it in concrete to prevent any further release of radioactive material (described in Chapter Three). There was also the problem of trapping airborne radioactive particles, and a special film of polystyrene was spread over soil around the site to prevent dust and contaminated particles from being stirred up into the atmosphere.

Once these most urgent local problems had been addressed, thoughts turned to the broader and longer term problem of decontaminating areas over which the radioactive fallout had landed. Many of the radioisotopes in the fallout (like caesium-137 and strontium-90) decay very slowly (see Table 2.3) and they are long lasting. They are also hazardous to humans (see Table 2.5). They could not be allowed to remain wherever they landed, and neither could they be left for the rain to wash away or wind to blow away (other areas would then be contaminated); the health risks were too great. Moreover, vast tracts of farmland would be sterilized and unfit for productive use for decades if they were not cleaned of dangerous radionuclides. The Soviet authorities had no option but to carry out an immense decontamination programme, without delay.

Soviet scientists used military geiger counters to plot detailed maps of contaminated areas in and around the Chernobyl complex. They also surveyed local forests and fields. Armed with this distribution information they could locate and retrieve radioactive pieces of graphite which had been blown from the core in the explosion, then load them onto (presumably covered) lorries for disposal. No details have been given, or emerged, about how or where the contaminated material was disposed of. The belief is that it was dropped down abandoned deep mines in geologically stable areas, where it would not be disturbed by earthquakes or disturb natural groundwater resources.

The location and clearance operation moved progressively outwards from the nuclear site. Work continued round the clock. The stakes were high - the longer areas remained contaminated, the greater the risk to human health. The overriding goal was to get the first two reactors at

Chernobyl (which were undamaged in the accident) ready to go back into operation with a minimum of delay. Each day without their electricity output was costing the Soviet economy, and placing heavy demands on the other power stations in the Ukraine.

Decontamination was a slow operation because it was dangerous. Personnel could only spend a few minutes at a time near the reactor, even wearing clumsy and heavy protective clothing. Special bulldozers were used with lead-shielded cabins to protect the drivers, and radio-controlled machines were used in the most heavily contaminated zones. Teams of volunteers were drawn from a wide area. Pay was good and risks (to their own health and to their future children) were high.

Soviet press reports speak of a gallant sense of duty and moral obligation amongst the workers who were prepared to make the necessary sacrifices. Initially they lived in caravans close to the reactor site, under cramped conditions with few home comforts. But the authorities recognized their needs, and by autumn 1986 they were moved to a special village which had been built for them a safe 50 km away (with shift transport provided).

Hundreds of acres of contaminated top-soil was scooped up, and this created serious problems of finding suitably accessible but safe sites to bury it. Soviet scientists were immediately put to work finding the best procedures for decontaminating affected soils. One possibility was to spray the ground with calcium compounds, water and phosphates. The hope was that these would seep into the soil and draw radioactive material below the plant root layer, effectively out of harm's way for crops and vegetation to re-grow.

The need to protect water supplies from contamination was of great importance. Within days of the accident work started on building embankments and check dams along the main local rivers, especially the Dnepr (which supplies around half of Kiev's drinking water), to prevent contaminated water from entering them and polluting the region's water supplies. The first phase of this work was said in the Soviet press to have been completed within two weeks.

The Soviet authorities took the threat of water pollution very seriously. Boris Paton, Chairman of the Ukrainian Academy of Sciences, reported at the end of May 1986 on plans to establish "artificial geochemical barriers" on the "migration paths of radionuclides in the groundwater". These were designed especially to save Pripyat River and the Kiev reservoir, after the base of the damaged reactor building had been reinforced with concrete (see Chapter Three).

But contingency plans were also being set in motion. For example people in Kiev were warned not to bathe in the Dnepr or spend time on the contaminated river beaches until further notice. An emergency water pumping station was built in a swamp near the Desna River to be used for water supplies if the Dnepr became contaminated. By early May bread and milk factories in Kiev had been warned not to use local river water and had started drawing water supplies from newly drilled artesian wells. But not even underground supplies were guaranteed, and by early June a number of wells in Byelorussia were found to be contaminated with radionuclides.

Once the embankments had been built on local rivers, efforts were started to decontaminate the whole area, starting with Pripyat and covering all local towns and villages. Huge amounts of contaminated soil and vegetation were removed and disposed of safely, roads and the outside of buildings were washed down to remove radioactive particles. The aim was to allow people to return home as soon as possible, although 18 months later most people were still in involuntary exile.

Contamination of livestock and agricultural land was a further major problem. Serious radioactive fallout had covered an area of around 1,000 sq km; fields were taken out of cultivation and work stopped in factories and on building sites. An estimated 10,000 cows were affected by the fallout, many so seriously that they had to be destroyed immediately. The rest were removed from the area for recovery, to allow radioactivity in their bodies and milk to decline naturally.

Dr Yuri Israel, Head of the Ukraine State Committee for Hydrometeorology and Environmental Control, announced early June that "the greater part" of the contaminated territory will eventually be "restored to the economy", although he offered no view on how long that might take. Farm produce and fields throughout the Ukraine and parts of Byelorussia are likely to be unusable for many years to come. Some western observers were expecting that the accident might trigger a new wave of Russian orders for (uncontaminated) pedigree cattle from the west (especially from Britain).

There was the further problem of fallout over forests. Most of the local fallout landed on the forests and wildlife of the Pole'ssye region. These are mainly managed conifer plantations and they appear to have been badly affected by the radioactive pollution. During early June 1986 forest fires in Pole'ssye started to recirculate radioactive dust and gave rise to further concern. Valerii Brezhnev, the Ukrainian Deputy Minister for Forests, said at the end of June that many plantations will have to be felled and the clearings ploughed up. He also revealed that veterinary monitoring of wild animals (elk, roe deer, boar and other species) had already begun in late May, and it would continue for a long time.

A big question mark remains over the question of just how badly the area around Chernobyl was contaminated. Western scientists find it hard to believe that it will be possible to fully decontaminate the whole area, certainly within years if not decades. Chernobyl's dark shadow over the Ukraine and Byelorussia will be there for some time to come.

Health

Many question marks also remain over the health risks to humans, both in the immediate danger zone and further downwind across the Ukraine and Byelorussia. The immediate toll of Chernobyl was 31 dead (mostly from severe radiation poisoning) and 200 hospitalized (mainly in Moscow) with variable degrees of radiation poisoning. Some 135,000 people were evacuated from their homes within 30 km of the power plant, to avoid continued exposure.

Many of the bone marrow transplant patients treated by Gale and his team (see Chapter Four) were simply beyond salvation, and there was considerable speculation over how many of the hospitalized and evacuated were doomed to die after exposure to Chernobyl fallout. Ukrainian press reports that victims of the accident will have their names commemorated in new streets in Chernobyl and Pripyat brought little comfort.

Radioactivity around Chernobyl rose sharply during the first few hours after the accident, and winds blew it over Pripyat. By 7.00 am on 27 April activity levels in Pripyat were between 180 and 600 millirem an hour, up to 50 times normal background levels. Levels continued to rise up to a peak of between 720 and 1,000 millirems an hour by 5.00 pm that same day. By 2.00 pm, according to the Soviet report (Chapter Seven), evacuation of the town's 45,000 residents was under way. The report estimates that people in Pripyat will have received between 1.5 and 5 rads of gamma radiation, between 10 and 20 rads of beta radiation, and perhaps up to 30 rads to the thyroid gland.

The critical group is likely to be the 135,000 exposed to high levels of radiation for up to three days before evacuation (see Chapter Four). Professor Leonid Ilyin, Vice President of the Soviet Academy of Medical Sciences, told a press conference in August 1986 that the Soviet authorities have established a register of each of these 135,000 people. Their actual external exposure to radiation had by then been checked, and their health would be monitored through their lifetimes. The official Soviet report (Chapter Seven) predicts a few hundred extra cancer deaths amongst this group over the next 70 years, compared with an expected death toll from natural causes of roughly 14,000.

But an estimated 100 million people living across the western Soviet Union and adjacent parts of Europe had also been exposed to fallout from Chernobyl. Most are believed to have received external exposure to fairly low levels of radiation, particularly of iodine-131 and caesium-137 (see Figure 5.4), equivalent to roughly a few years' natural background radiation. It is also believed that over the next 50 years total additional external exposure from the accident will be equivalent to less than 10% of the natural background radiation over the same period.

As a rough estimate the Soviet report (Chapter Seven) calculates that external exposure to the relatively short-lived radionuclides from Chernobyl is likely to increase cancer mortality by around 0.5% (ie 1 in 200) in western Russia. This translates into around 5,000 additional deaths. Set alongside the 9.5 million 'normal' deaths from cancer expected within this population over the same timescale, the impact of Chernobyl might turn out to be relatively small.

But people were not simply exposed to external radiation from the accident. There are other possible pathways of exposure (see Figure 2.2), the most important being internal radiation from eating and drinking contaminated products. The two key isotopes are iodine-131 (which causes short-term damage) and caesium-137 (long-term). The health risks really depend on what people ate and drank at the time. Many Russians are believed to have consumed dairy products with radiation over 200 times the 'safe' limit for iodine-131. The Soviet report concludes that some will have received thyroid doses reaching hundreds of rad, and estimates a possible 1% increase in death rates from thyroid cancer in affected area over the next three decades. This would amount to an additional 1,500 deaths.

The Russians anticipate that the most serious long-term health threat is likely to be the caesium-137. Soils across the main fallout zone are very poor in organic matter, and caesium uptake by plants may be between 10 and 100 greater than in soils elsewhere. Crops and dairy produce from this area would be highly contaminated. Preliminary estimates in their report suggest a likely increase of 0.4% in cancer deaths in the western USSR, causing a premature loss of some 40,000 lives over the next 70 years.

Dr Morris Rosen of IAEA, speaking at a press conference in Vienna on 26 August 1986, described the Soviet estimated death toll as "conservative". Using risk factors recommended by the International Committee on Radiological Protection (ICRP), he estimated that external exposure would produce more than 24,000 additional cancer deaths (not 5,000) over 70 years.

Whichever figure is used (and only time will tell what the real damage to human health is), the estimates are relatively low compared with other major causes of premature death. For instance, roughly 20,000 deaths from medical uses of radiation (for diagnosis and therapy) might be expected in a population of 100 million over 70 years; in the order of 100,000 can expect to die from exposure to natural sources of radiation. Road traffic accidents, Aids and other health risks are likely to have a much greater impact than fallout from Chernobyl.

Of course this does not *excuse* what happened, or make it more acceptable. Neither does it make it any easier for the countless thousands who are doomed to die because of Chernobyl. Theirs will be a lingering, debilitating death. The sinister shadow of Chernobyl will close in on them progressively more and more through time, until it eventually takes their last breath.

Soviet energy supplies

It was clear to observers inside and outside of the Soviet Union that the accident would inevitably mean a short-fall, however temporary, in electricity production. The four nuclear reactors at Chernobyl were providing 15% of the country's nuclear power, which was much needed for industry.

This problem turned out to be less serious than expected. The short-fall in power generation was averted by running all other power stations in the Ukraine at full capacity. This was very much a stop-gap measure which could not be continued indefinitely, and there was considerable pressure to get the two undamaged reactors at Chernobyl back on-stream as soon as possible. The Russians coped with the loss of power very well indeed, and many economists in the west now believe that it will prove to be impossible to detect any impact of Chernobyl in statistical reports on energy.

There were also fears, especially within the Soviet Union, that the accident might be a serious blow Soviet plans to expand their nuclear programme. These turn out to have been unfounded (see Chapter Ten).

Soviet economy

The Soviet Union is never very forthcoming about aspects of the economy which it would prefer not to have examined in public, so it is difficult to speculate on the possible economic costs of Chernobyl. They have estimated that the accident probably cost them around 2,000 thousand million roubles in 1986 alone, but there is no break-down of this.

Some cost headings are fairly self-evident, even if precise figures are not available. Enormous levels of resources were invested in the immediate struggle for containment on site, and the massive decontamination programme through the region. The medical costs must also be vast, given the large numbers involved and the inherently costly nature of cancer prevention and treatment. Rehabilitation costs for the evacuees must also be high, given the numbers involved and the need to provide them with homes, furniture and other basic necessities.

Early fears that the whole Soviet economy might be adversely affected by the accident have turned out to be largely groundless, because electricity production was not seriously diminished.

Perhaps the biggest immediate problem was the prospect of serious and lasting damage to agriculture in the western Soviet Union. The Ukraine is Russia's *'bread-basket'*. It has traditionally produced vast quantities of crops, meat, milk and cheese for consumption throughout the country. It is almost inevitable that crops and products from this area will be unsaleable (on health grounds) for many years, either until decontamination is complete or activity levels in the radionuclides present falls naturally to an acceptable level.

This previously productive farming heartland is likely to be a sterile wasteland unsuitable for raising cattle and other livestock, growing fruit, vegetables and winter grains (like wheat and barley), and producing dairy products. This will inevitably be a serious blow to the local economy. But whether it has a serious and lasting effect on the national Soviet economy, possibly further aggravating the common problem of food shortages and encouraging greater imports of foodstuffs, remains to be seen.

Planecon, a US research group based in Washington DC that studies the Soviet Bloc countries, has estimated that Chernobyl might cost the Soviet Union up to $4.3 billion in medical expenses and losses in agriculture and other sectors of the economy. This is at best only an informed guess, but it does highlight the possible size of the economic problem for the Soviet Union.

Soviet credibility

Observers in the west also believe that there could be political problems as a result not so much of Chernobyl itself, but more of how the Soviet authorities handled the accident. This is likely to cast a shadow over international credibility of Mikhail Gorbachev's attempts to show a new style of Soviet leadership - the so-called *glasnost*, founded on responsible and open attitudes.

Soviet politicians, including Gorbachev, face an uphill struggle in convincing the west that they acted properly over Chernobyl. Many western politicians were incensed over what some saw as Soviet unwillingness and others as inability to inform people (at home and abroad) what had happened and how they were tackling it. For example, Gorbachev first spoke in public about the accident on 14 May, after what *The Times* called "18 days of deafening and ill-judged silence".

During those 18 days large numbers of Russians were left in the dark, with no advice given on what they should do to minimize health risks. The exodus from Kiev (see Chapter Four) was borne of panic and lack of guidance, more than real need. Similarly neighbouring countries were given no information which might have helped them make decisions about coping with the radiation cloud. Poland, for example, had to make informed guesses about the use of prophylactive iodine (see Chapter Four) which would have been easier and more reliable had they been better informed about levels of radioactivity in the cloud and the radionuclides involved.

The Soviet image and credibility were badly dented as the Russian authorities came under criticism from other countries on two grounds. One was their competence to handle a major disaster properly. The other, and in the long term potentially the more damaging, was their honesty. This latter spilled over into the international strategic debate, as western leaders asked how can the Russians be trusted to be honest and open over arms control (where verification is notoriously difficult) when they were being so patently unforthcoming over Chernobyl (where some verification was possible).

The key question is why the Soviet authorities made no immediate announcement that the accident had happened, and only admitted it after repeated questioning from Sweden. The cynical answer is that the Soviet leadership knew about the accident, but chose to remain silent. But some western observers now believe that relevant information was not passed to the leadership straight away, and it then proved difficult to get reliable information quickly and continuously. If this interpretation is correct, then the leaders were at fault for not ensuring that reliable information channels were properly managed; they were not necessarily guilty of deliberate and calculated deception, as some have suggested.

The other important question is why did the Soviet authorities not make stronger efforts to keep other countries informed as they pieced together what had happened. Why, for example, did they not release information on the quantity of radiation which had leaked out, and on the radio-nuclides concerned? Other countries could have reacted better and quicker to minimize health risk to their own citizens if they were armed with that type of information, rather than having to rely on informed guesswork. One explanation is that the Soviet authorities wanted to suppress that sort of information for as long as they could, to avoid the risk of widespread public panic. They must also, and inevitably, have directed most of their resources, time and man-power towards dealing with the problem rather than talking about it.

Besides, the Russians might legitimately claim, Britain and the United States have proven that they are far from forthcoming with information about nuclear accidents, particularly close to the event. It took an inexcusable 27 years before fuller (but far from full) details of the 1957 graphite fire at Windscale (Sellafield) came to light (see Chapter One), and then only after repeated probing by anti-nuclear pressure groups. Little reliable information was forthcoming at the time of the Three Mile Island incident.

The Russians have also been strongly accused of mishandling the evacuation of locals from around the site (see Chapter Four). But many of the critics overlook the vast nature of the problem. Mobilizing 1,000 buses with drivers, briefing marshalls, advising the evacuees on what to take and what to leave, deciding where to take them, arranging the provision of housing and food - co-ordination of such an exercise takes time, even in a centrally controlled Soviet system and given customary Soviet military-like organizational strengths.

There are many in the west who would complement the Russians on the speed and efficiency of their evacuation of such massive numbers of people. They would seriously question whether western countries could have done anywhere near as well, let alone better. The evidence from Three Mile Island, for example, indicates that the agreed evacuation procedures were largely ignored (granted the evacuation was voluntary, not enforced). There are grave doubts whether the evacuation of 135,000 people could be handled speedily and efficiently in Britain or the United States.

The issue of Soviet credibility is thus a somewhat mixed one. As the accident was happening, and as radiation was spilling out and spreading over Europe, there was a cacophany of cries from world leaders about Soviet handling of the whole affair. With the passage of time, however, and as more considered judgements have emerged, the criticisms have evolved into sympathy over the immensity of the problems they faced.

In the long term it is doubtful whether Chernobyl will have a lasting impact on Soviet credibility. It will progressively be eclipsed by other issues of international importance, and *glasnost* will replace traditional Soviet secrecy. This long shadow of Chernobyl will soon be lifted.

LONG-TERM PROBLEMS ELSEWHERE

The most serious long-term problems from Chernobyl will be concentrated within the Soviet Union. But they will not entirely be confined there. Radioactive fallout from the accident covered many countries, and the total cost of damage will only emerge with the passage of time. By far the biggest problems will be contamination and damage to health.

Contamination

Whilst most countries in Europe received some fallout from the accident (see Figure 5.2), levels and patterns of deposition varied considerably across the continent. Outside the Soviet Union most fallout landed over Switzerland and parts of Austria, northern Italy and southern West Germany (see Figure 5.4). Despite the high levels of activity over Sweden and Poland during the first few days, these countries are believed to

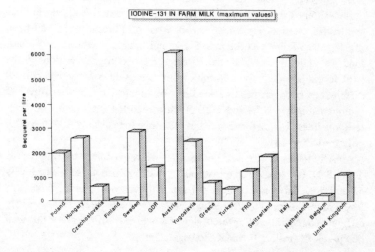

Figure 8.1 *Maximum levels of iodine-131 in farm milk in different countries, May 1986. After Salo (1986)*

Table 8.1 Summary of radiation levels found in member states of the European Community after Chernobyl

country	soil (Bq m^{-2})		milk (Bq l^{-1})		leafy vegetables (Bq kg^{-1})	
	I-131	Cs-137	I-131	Cs-137	I-131	Cs-137
Belgium	1,700		150-340	nd	1,000	nd
Denmark	35-1,040	5-300		15-34	500	140
West Germany	10,000-120,000	nd	40-1,200	nd	44-3,000	nd
France	32-960	40-500	40-500	40-100	400-3,000	40-60
Greece	nd	nd	350	100	2,000	109
Ireland	4		nd	nd	nd	nd
Italy	nd	nd	45-260	9-27	480-1,700	96-340
Luxembourg	nd	nd	250	nd	500-1,000	nd
Netherlands	1,000-10,000	nd	200	nd	800	nd
Portugal	11		nd	nd	nd	nd
Spain	nd	nd	65	40	200	
United Kingdom	9,000	5,000	40-480	200	50-100	40-803

nd = no data

SOURCE: Loprieno (1986)

have received high absolute but relatively low relative amounts overall.

It was only a matter of time before the fallout started to appear in fresh foodstuffs, particularly milk. Across most of Europe cattle had been grazing out-of-doors as the radiation cloud floated overhead, and grassland was quickly contaminated. There are pronounced variations in the peak levels of radioactivity in milk from country to country, and curious differences between relative loadings of iodine-131 and caesium-137 (compare Figures 8.1 and 8.2). Austria reported high peaks on both radionuclides, Switzerland and the UK only for caesium-137, Sweden only for iodine-131.

Activity levels in soils and produce varied from country to country (see Table 8.1), but so did interpretations of the significance of given levels of activity. Different countries have different acceptable maximum levels of radiation, or *'intervention levels'* (see Table 8.2). Consequently different countries have different points at which restrictions are imposed on the sale of fresh milk, meat and vegetables.

Health

There is inevitably considerable interest outside Russia in the question of

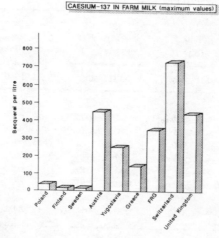

Figure 8.2 *Maximum levels of caesium-137 in farm milk in different countries, May 1986. After Salo (1986)*

how much harm was done to the large numbers of people across Europe who were exposed to fallout from Chernobyl. They might have received doses of radiation which were small in absolute terms but relatively large in relation to normal background levels.

People in Europe were exposed to radiation from Chernobyl in various ways, which reflect the main pathways for radionuclides in the environment (see Figure 2.2). The four most important, in rank order, were:

(1) direct irradiation as the radioactive cloud passed overhead,
(2) inhalation of radioactive substances from the cloud,
(3) beta ray contamination of skin, and
(4) external irradiation from deposition on soil and the consumption of contaminated food and drink.

Western scientists have tried to estimate what levels of exposure would have been received by people living in different places. Some of the estimates are based on direct measurements of fallout (see Figure 8.3), and others are based on simulation modelling (see Figure 8.4). Different studies tend to give slightly different results, which makes it difficult to arrive at general conclusions. The simulation results suggest that people received high doses in Poland, Switzerland, southern West Germany,

Table 8.2 Examples of recommended maximum levels of radiation adopted in certain foodstuffs

country	milk (Bq l $^{-1}$)		vegetables (Bq kg $^{-1}$)		meat (Bq kg $^{-1}$)	
	I-131	other	I-131	other	I-131	other
European Community	nd	370 Bq.kg^1	nd	600	nd	nd
Belgium	nd	500	nd	1,000	nd	nd
France	nd	3,700	2,000	nd	8,000	nd
Germany	nd	500	250	100	200	100
Luxembourg	500	nd	250	nd	nd	nd
Netherlands	500	nd	1,300	nd	nd	nd
United Kingdom	2,000	nd	nd	nd	nd	nd

nd = no data

SOURCE: summarized from Loprieno (1986)

northern Italy and northern England (Figure 8.4). Reported exposure rates were particularly high in Sweden, Finland and Yugoslavia (Figure 8.3).

Only time will tell whether or not people outside the Soviet Union are likely to suffer acute health problems after Chernobyl. The casualties will be long-term, probably relatively small in overall number and inevitably widely scattered in distribution. This means that it will be extremely difficult if not impossible to detect this group amongst the larger group who are unfortunately likely to develop cancer anyway over coming decades.

Some experts predict a total number of additional cancer deaths from Chernobyl in the region of 1,000 across the whole of Europe. But this is a summary statistic, and precisely who these doomed individuals are is impossible to say at present. However, whether they can be identified as a separate group or not is largely an academic question, because it will not materially alter their long-term survival prospects one way or another.

Cost

Cutting across the problems of contamination and damage to health is the overall question of the total cost of Chernobyl. There are two areas of interest - what will the total cost be, and who will pay it?

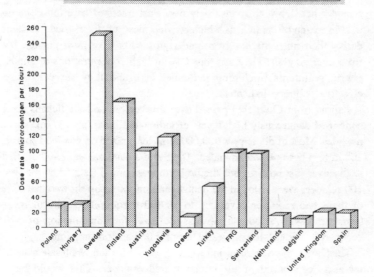

Figure 8.3   Exposure rates reported to IAEA, 4 May 1986. After Salo (1986)

Total cost might be measured in a variety of ways. Loss of public faith in nuclear power (see Chapter Ten) is a cost, both in terms of lost opportunities and in hard cash terms. What is the cost of the massive image building campaigns which the British nuclear industry, for example, launched in the wake of the accident?

In monetary terms, the damage bill should cover health care, decontamination, compensation to farmers for unsaleable contaminated produce, and so on. It will be interesting to see whether any detailed breakdowns of total costs appear in the future; none were available 18 months after the accident. The only figures available are crude estimates. For instance, Sweden put the cost of food by fallout at around $144 million in 1986, and the West German Government believes its financial losses (from confiscation and destruction of contaminated foodstuffs and restriction of food production in high fallout areas) run to several hundred million marks.

Two areas where there is particularly keen interest in Chernobyl are the upland sheep farmers in Britain (see Chapter Six) and the Lapps in Scandinavia.

Chernobyl could have a catastrophic impact on the 7,000 or so nomadic

Lapps (of a total Lapp population of around 70,000) of northern Scandinavia. The Lapp economy is heavily dependent on the success of their reindeer herds, which roam freely over vast tracts of bare inhospitable land in search of the lichens which provide most of their food. Lichens derive their nutrients not from underlying soils (like most plants), but from overlying air. They can thus take in high concentrations of atmospheric pollutants (including radionuclides), and they serve as very effective pollution indicators.

Fallout from Chernobyl passed over this region, and the lichens soon contained dangerously high levels of iodine-131, caesium-137 and other nuclides. Most of the Lapp's 900,000 reindeer grazed on this food source, and in turn became contaminated. The Swedish authorities were horrified to discover just how serious the problem was. Initial tests when the first 100 reindeer were killed in the annual autumn round-up showed that 97% of them had radiation levels up to 10,000 becquerels per kg of meat (compared with Sweden's permissible level for consumption of 300 becquerels per kg). It was clearly unfit for human consumption.

The Swedish government immediately called for the wholesale slaughter and deep burial of the contaminated reindeer. This would have instantly wiped out the Lapp's resource base and inevitably led to the

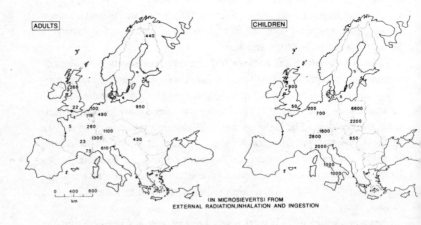

Figure 8.4   Total effective dose equivalents across Europe. Based on estimates of total effective dose equivalents from external radiation, inhalation and ingestion compiled by the World Health Organization. After Anon (1986) IAEA Bulletin (Autumn)

entire collapse of their culture. The price was too high! Instead the government decided to pay the Lapps the normal price for every reindeer killed, and then release some of the meat to feed mink and foxes in the country's fur farms.

The immediate problem was 'solved', admittedly in a way that the Lapps had to tolerate rather than welcome. But the long-term problem remains. How can this culture survive when the feeding grounds for their stock are so badly contaminated with fallout, their reindeer herds are so badly depleted, and there is no large market for their produce? This particular shadow of Chernobyl could be particularly long and sinister.

The ultimate question facing all parties who have suffered as a result of Chernobyl is "Who will foot the bill?" Logically the Soviet Union should offer to meet the compensation costs incurred by governments across Europe. On 16 May 1986 the European Parliament urged EEC ministers to evaluate the damage caused by the accident and then present Moscow with a bill. There are no reports that any country had done this within the following two years, and similarly no signs that the Russians were volunteering to pay.

LESSONS OF CHERNOBYL

There are many important lessons to be learned from Chernobyl. Some are already clear and others will emerge in the course of time. The fact that the accident happened on Soviet soil dictated much about how the immediate problems were handled. The nuclear industry and many politicians world-wide were quick to isolate peculiarities of the Soviet system, which they insisted, had allowed the accident to happen in the first place.

But the lessons are there for the whole world. Many groups of people must learn from the experience of Chernobyl - the public, politicians, nuclear engineers, scientists, international agencies, those involved in radiological protection, and so on - both inside and beyond the Soviet Union.

Some of the lessons are entirely technical, and they lie within the realms of nuclear engineering and reactor physics. The whole nuclear industry world-wide will have to think carefully about the broader implications of reactor design, construction, containment, personnel. These are dealt with in Chapter Nine.

Other lessons centre on the impact of the accident on public feelings and fears. Governments and the nuclear industry around the world have been presented with much firm evidence of how public confidence in nuclear safety and faith in nuclear power can be rocked by a major radiological accident (see Chapter Ten). They would do well to reflect on that evidence, particularly in light of our growing dependence on nuclear power.

Yet other lessons are scientific, and they will emerge through time as the results appear of the unique opportunity to learn more about the biological effects of radiation and about patterns, rates and processes of transport and uptake of radionuclides in the environment. Early studies of Chernobyl fallout over Britain, for example, have uncovered several interesting findings. Thunderstorms appear to be particularly effective in washing radionuclides out of the atmosphere (see Figure 6.2). Scientists have also been puzzled by the unexpected 'stickiness' of radionuclides in some upland soils (ie they have not been washed out as quickly as theory might have predicted).

There are a number of critical areas where operational lessons must be learned.

Radiation monitoring

The accident provided a golden opportunity for countries to test the reliability and effectiveness of their radiation monitoring systems, installed to detect military fallout as well as civil nuclear accidents. Resolution and reliability of monitoring in some countries was proven to be very good. For example, Sweden's sensitive detectors picked up traces of the radiation cloud much earlier than it was found anywhere else (see Chapter One), and Poland's monitors provided them with useful early warning that the cloud was passing overhead (see Chapter Four). But some countries found their networks to be less than perfect. In Britain, for example, the problem of gaps in coverage around the country was a serious one (see Chapter Six) which the Government resolved subsequently to remedy.

Radiological protection

Chernobyl established the need for greater harmonization of so-called *'intervention levels'*, or *Derived Emergency Reference Levels* (DERLs) of radiation doses. These are the recommended upper levels of radiation in foodstuffs, and the levels adopted vary considerably from country to country (Table 8.2). Consequently different countries had different reactions to the Chernobyl fallout (see Chapters Four to Six), depending on whether activity levels (Table 8.3) exceeded their national intervention levels or not.

There is no *absolutely* safe level of radiation exposure, other than zero. Even natural background levels of radiation (see Chapter Two) pose some risk to humans, but it is usually negligible. The impacts on health of high levels of radiation (particularly from external exposure) are known with some certainty (see Table 2.4). But scientists still know relatively little about the long-term biological effects of exposure to small doses of radiation, and this is where the problem lies.

Intervention levels for individual radioisotopes are determined primarily on scientific criteria. But the scientific evidence does not define critical absolutes at low levels of exposure. Precisely where the intervention levels are fixed is decided in the political arena on the basis of acceptability of risk.

Some countries tolerate higher radiation levels in foodstuffs than others. But this should not be so. Human susceptibility to radiation poisoning does not vary systematically from one country to another! The intervention levels need to be standardized (preferably at the lowest common level for a given radioisotope). This would give wider protection to the public, but it would also help to minimize international political tensions come the next major radiological emergency involving trans-frontier contamination and trade in foodstuffs.

Had such harmonization occurred pre-Chernobyl, the time-wasting and acrimonious dialogue over a European ban on food imports from eastern Europe (see Chapter Five) could have been avoided and food markets in the west would not have contained contaminated produce.

Availability of information

Availability of reliable information about radiation and the risks to health, in a form that the public can understand, also emerged as a critical issue during and after the accident. An uninformed public is often a scared public, so the argument (sometimes used by politicians and within the nuclear industry) that it is better to tell people little to avoid panic reactions is difficult to defend. Everyone involved in dealing with major or minor nuclear accidents must learn how and when to provide the type of information that people need. Glossy publicity leaflets and colour videos which paint the picture of nuclear power as safe, reliable and problem-free (see Chapter Ten) may be of general interest to the public, but they are no substitute for honest advice and open reporting about nuclear operations and safety.

Table 8.3 Comparison of risks - nuclear and others

cause of death	probability of death per year
cigarette smoking (20 per day)	1 in 200
accidents in deep sea fishing	1 in 400
natural causes, 40 years old	1 in 500
accidents on the road	1 in 5,000
accidents in the home	1 in 10.000
accidents at work	1 in 20,000
air crash	1 in 100,000
radiation from nuclear facilities (1,000 µSv per year)	1 in 100,000
from possible accidents due to every 100 nuclear power stations operating	1 in 5,000,000,000

SOURCE: Anon (1986) *Nuclear India* 24; 5

Risk assessment

There are also lessons for risk assessment. To many people nuclear installations are inherently dangerous, and should be dismantled without delay. But their thinking is founded on fear of the unthinkable, rather than on sober consideration of the evidence. The problem is that risk assessments *must* be based on experience to date, and they should therefore always be open to revision. One major incident can radically alter statistical probability estimates, particularly if existing estimates are based on no previous incidents (ie they are largely academic).

The relevance of this to Chernobyl is more than passing. Before the accident, because there had been so few major nuclear accidents around the world, the estimated risk of people being killed by such accidents was regarded as small. Some comparisons of different types of risk (pre-Chernobyl) are shown in Table 8.3.

This type of evidence is used by those who argue that "nuclear power is safe, because you stand a much bigger chance of dying in a car crash or in an accident at home or work than from radiation poisoning". That might well be true in purely statistical terms, but an individual can minimize his or her exposure to most of the other risks if they so choose. The shadow of Chernobyl respects no-one. Furthermore, the other risks are likely to affect people throughout a country, whereas the nuclear risk is spatially concentrated. Moreover, nuclear accidents can claim many victims in one go, whereas the other risks don't tend to.

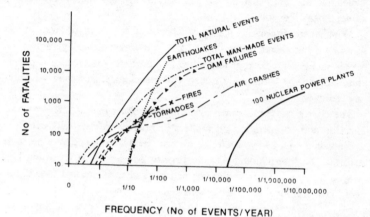

Figure 8.5 Death tolls of various natural and man-made events. After Eisenbud (1979)

Nuclear risks (based on probability estimates) are also minimal when viewed alongside other major hazards like earthquakes, dam failure or major fires (Figure 8.5). But these other hazards are more common. One or two Chernobyls would radically alter the risk assessments for nuclear power, and might erode the complacency which some people have in countries like Britain (see Chapter Ten).

INTERNATIONAL CO-OPERATION

Perhaps the biggest lessons from Chernobyl lie within the field of international co-operation. The Soviet authorities were heavily criticized for their handling of the whole affair, but this was completely uncharted territory!

Coping with major disasters is difficult enough at the best of times, but when a disaster has impacts which transcend national frontiers then basic aspects of international morality and law come into play. States *cannot* simply do as they please. Their actions should observe three duties

139

enshrined in international law. The first is *"due diligence"*, which requires them to make all reasonable efforts to prevent the pollution of areas beyond their jurisdiction, and combat it when it occurs. Next is *"good neighbourliness"*, which means they are not to allow their territory to be used in way that is prejudicial to the rights of other states. Finally there is *"solidarity"*, under which they must take the interests of other states into account when engaging in or allowing activities that adversely affect those other states.

Whether the Russian handling of Chernobyl failed on all three counts is a largely academic question, best left to international lawyers. But the whole incident revealed gaping holes in the existing international procedures for dealing with major trans-frontier accidents, particularly involving radiation.

There were many calls after Chernobyl for better international communication and co-operation on matters of nuclear safety. Mikhail Gorbachev said on 14 May 1986 that he was ready to co-operate in such efforts. The International Atomic Energy Agency (IAEA), which had played a key role in co-ordinating activities after Chernobyl (see Chapters Three and Five) was the obvious vehicle for overseeing this.

World leaders looked towards IAEA to operate more effectively as an international nuclear information centre. They wanted it to become the focal point in a world-wide system for early notification, warning and assessment of the potential environmental consequences of radiological accidents. They also demanded from it an ability to process and co-ordinate requests for assistance (in order to increase national capabilities to handle nuclear accidents), and to provide immediate consultations by top-level experts in safety and radiological protection.

IAEA experience

IAEA had won its battle colours at Chernobyl. Governments around the world, eager to find out what was happening there but frustrated by the prolonged Soviet silence, naturally turned to the Agency for advice. The initial requests (on Monday 28 April) centred on whether there had been a nuclear reactor accident, and if so where. After the Russians confirmed that it had happened at Chernobyl, IAEA was besieged with requests for technical details about the RBMK reactor and about the accident. Governments urgently needed to know how much radiation had been released, where it was being dispersed, and what radionuclides it contained.

The Agency was forced to act as a clearing house for information, but its hands were quite firmly tied. It was (and is) not empowered to officially request that member states report their radiation measurements. Moreover it appreciated that the relevant institutions would be hard at work collecting detailed radiation measurements for their own national authorities, so their resources would be fully stretched. Neither is IAEA authorized to advise member states what to do about public health matters.

Its only option, therefore, was to establish informal contacts with the radiation protection authorities around Europe, to build up the best picture it could about the size of the affected area. This it did in the last few days of April. Many individual countries volunteered information (see Table 5.1) to the Agency on dose rates in the environment, radio-isotopes detected in air, water, ground, grass and foodstuffs, iodine measurements in thyroids and results of whole body radiation counting.

IAEA also served a very important role in the immediate post-Chernobyl period. For instance, it ensured a reasonably regular and frequent flow of information after 26 April (see, for example, Figure 4.1). It later persuaded member states to give thought to steps required to keep one another advised about risks of trans-frontier pollution. It also arranged meetings with the Soviet authorities and confirmed to the world at large what had happened at Chernobyl and what the Russians were doing to cope with it and contain the problem (see Chapter Three).

Practical lessons

The Agency learned several critical lessons from Chernobyl. One was that information on the level and nature of radiation release after an accident *must* be made available, quickly, and in a form that is compatible between countries (eg using standard measurement units). It also found that existing guidance on how to cope with nuclear accidents was entirely inappropriate to this type of situation. For example, guidance procedures established by the International Commission on Radiological Protection (ICRP), the World Health Organization (WHO) and the IAEA were shown to be deficient in two ways - they aim mainly to avoid acute and harmful exposure to individuals, and they are relevant only to the area immediately surrounding the accident site.

The Chernobyl experience was of a much bigger dimension, and full adoption of the recommended procedures would have been prohibitively costly and difficult to implement. What's more, the disaster was so big and wide-ranging in its impacts that it had to be handled by many authorities, including some who are not normally involved in (and

141

generally not familiar with) radiation protection.

New conventions

IAEA emerged from Chernobyl anxious that it did not seem to be possible, under existing arrangements, to deal effectively with that type and level of trans-frontier nuclear problem in Europe. It saw a great need to develop a long-term co-operative programme involving the major international agencies. These include the World Meteorological Organization (WMO) for predicting likely distribution of fallout, the United Nations Scientific Committee on the Effects of Atomic Radiation (UNSCEAR) for assessing possible overall health consequences, the World Health Organization (WHO) for providing guidance to national health authorities on protection of health, the UN Food and Agriculture

Table 8.4 Summary of the IAEA Early Notification Convention

* covers all uncontrolled releases of radioactive material from any source, ir-respective of its nature and location, that may result in transboundary effects which could affect the radiological safety of another State (including all nuclear accidents on land, at sea or in outer space; excluding accidents connected with nuclear weapons and nuclear weapons testing, which might be notified voluntarily)

* if such an accident occurred, a State Party is required to notify immediately (directly, or via IAEA) other States which may be physically affected, and the IAEA, of (a) the nature of the accident, (b) the time of occurrence, and (c) the exact location, where appropriate. It must also provide them promptly with available information relevant to minimizing radiological consequences in the affected countries

* the Convention details what information must be provided by the notifying State, and requires that State to respond promptly to a request for additional information by an affected State Party

* IAEA would be the clearing house for receiving notification of a nuclear ac-cident and for providing relevant information to State Parties, Member States and appropriate international organizations.

SOURCE: summarized from Anon (1986) *IAEA Bulletin* (Autumn); 63-4

Organization (FAO) for guidance on necessary changes in food processing and farming practices, and the International Labour Organization
(ILO) for guidance on the protection of workers in contaminated sur-
roundings.

The Agency's initiatives were revealed at the meeting of governmental
experts it arranged in Vienna (21 July to 15 August 1986). Some 286
delegates from 62 Member States and 10 international organizations
were present to hear the Russians present their formal report on Chernobyl (see Chapter Seven).

Two new international agreements on nuclear safety, designed to plug
critical gaps in existing procedure, had been drafted by experts and were
adopted by consensus at the final plenary session of the meeting (on 15
August). The *Convention on Early Notification* (Table 8.4) requires

*Table 8.5 Summary of the IAEA Convention on Assistance in the Case
of a Nuclear Accident or Radiological Emergency*

---

* the aim is to establish an international system to facilitate prompt provision of
relevant assistance, directly among State Parties or through IAEA, and from it and
other international organizations

* State Parties are required to notify IAEA of experts, equipment and materials
they could make available for the provision of emergency assistance to other
States

* the requesting State would assume responsibility for the overall direction and
control of the assistance, and provide support services and effective administra-
tion. This would include

  - indemnifying the assisting Party against claims from third parties
  - providing transit through their territories of personnel, equipment
  and property involved in the emergency assistance
  - granting the assisting personnel appropriate privileges and immunities
  to carry out their functions

* IAEA would play a key role in facilitating and supporting the co-operation
amongst State Parties in emergency assistance (eg in maintaining liaison with
other international organizations, providing expert services and manpower training and development)

SOURCE: summarized from Anon (1986) *IAEA Bulletin* (Autumn); 64

signatories to report - either through IAEA or directly to the nations that might be affected - the location, radiation release, time and other relevant data on all nuclear accidents (other than those relating to weapons) that may have trans-boundary consequences. The *Convention on Emergency Assistance in the Case of a Nuclear Accident or Radiological Emergency* (Table 8.5) sets up an international network for providing experts, equipment and other materials that might be needed in the event of a major accident. Fifty-one nations (including the Soviet Union) signed the two conventions, and they became effective on 27 October 1986.

Internal changes

IAEA also seized the opportunity of introducing changes to its own practices, having discovered some deficiencies as a result of the Chernobyl experience.

For instance, it set up a new ASSET (Assessment of Safety Significant Events Team) group to carry out detailed analyses of the operational experience of nuclear power plants in the field of safety. One of their briefs was to use experience of past accidents to develop more effective preventive measures. The Agency also expanded its Incident Reporting System (IRS) to include more significant events and introduce more effective mechanisms for analysing reported events.

There are also plans to enlarge the Agency's emergency response mechanisms by establishing an Emergency Response Unit to process and co-ordinate requests for assistance and communications. It is also planned to expand the Radiation Protection Advisory Team (RAPAT), in order to help non-nuclear power member states to establish a radiation protection capability.

IAEA played a leading role in the Chernobyl drama, for which it deserved and won widespread acclaim. But it took its position seriously, and drew its own conclusions about its effectiveness. It drew its own lessons, too, about the need for greater international co-operation and dialogue. Many experts agree that, whilst the two new conventions will not *prevent* an accident like Chernobyl from happening again, they should at least help countries to cope with it more effectively and lessen the risk of serious contamination and large-scale human suffering .

CONCLUSIONS

Many people in the west believe that the Russians handled the accident at Chernobyl remarkably well, given the almost impossibly difficult conditions they were faced with. Academician Vevgeny Velikhov (of the Soviet investigating committee on Chernobyl) said at end of May 1986 that they confronted unprecedented problems - there was little advance planning or understanding (anywhere in the world) of how to manage such a badly damaged reactor, and the information that was available was often contradictory.

In many ways it is doubtful whether any western country could have done much better. Much of the western political and diplomatic anger at the Soviet Union over its handling of Chernobyl at the time reflected exploitation of a superb propaganda opportunity; here, after all, was a great Super Power struggling to hide a major disaster, contain its impacts and prevent public panic.

Dr Hans Blix told IAEA's Board of Governors on 21 May 1986 that "it is clear that there is a will in the whole world to draw all the lessons we can from the accident, and to reduce further the risk that any such accident would occur again. There is an equally strong interest to ensure that if, in spite of all precautions, an accident were to occur, the consequences of it would be limited".

Chernobyl was widely described at the time as "the world's worst accident". The statement, whilst well intended, might not be entirely true; it depends on the yardstick of 'badness' which is used. If *immediate* death toll is used, then the 31 Russians who died from Chernobyl are greatly over-shadowed by the 1,500 people who died when the Titanic sank.

It was certainly the world's worst *nuclear* accident. It now appears to have been less serious than was first thought. Early western media coverage was speculating that thousands *must* have died straight away, and we now know that this simply wasn't the case (see Chapter Four). Of course the total death toll will be much higher, but exactly how much higher is a matter of debate.

It was without doubt a highly unusual event, and hopefully it will turn out to have been a unique one. But 'Murphy's Law' ("if something *can* go wrong, the chances are it *will*") suggests that it is likely not to be. If that is so, then the sooner the lessons of Chernobyl are learned, the better.

Given the world's growing reliance on nuclear power (see Chapter Ten), it might only be a matter of time before the next 'Chernobyl'. If there is better radiation monitoring, better radiological protection, better availability of information and better internatinal co-operation then, whilst the next Chernobyl will be no more pleasant an affair, it should be a safer one all round.

9
IT CAN'T HAPPEN HERE ...

The western television news coverage of Chernobyl during late April and May 1986 was thorough and graphic. Here was a catastrophe of global proportions unfolding before our very eyes. Many people were inevitably concerned about where the radioactive fallout would land and what the risks to human health (including their own) were likely to be. Many reports stressed that this was the worst accident in the history of nuclear power. But history has a habit of repeating itself - witness the history of near misses in the nuclear industry's track record (see Chapter One). Since the accident, the name 'Chernobyl' has been used as more or less synonymously with 'nuclear accident', regardless of the nature and magnitude of accident.

Inevitably the question of whether similar accidents happen might again in the future, in the Soviet Union or perhaps closer to home, came into many people's thoughts. The anxieties were fuelled directly by what appeared to be happening at Chernobyl. But other factors were also relevant, including the increasing number of nuclear power stations around the world, especially in developing countries, and the growing reliance of major nations like Britain on nuclear power. Just because Chernobyl *should not* happen again does not mean that it *will not*!

Soviet nuclear industry

Nuclear specialists in the west, eager to distance themselves and their industries from the dark shadow of suspicion that Chernobyl *could* happen elsewhere, were quick to draw their own conclusions about why the accident had happened. They pursued two main themes, designed to throw doubt on the reliability of the entire Soviet nuclear power programme and show that the nuclear industry in the west is more responsible, more reliable and inherently safer. One was to argue persistently that the Soviets employ old nuclear technologies. Around half of the existing Soviet nuclear reactors are graphite moderated, whereas allegedly safer water-moderated reactors are favoured in the United States.

The second main argument was that Soviet nuclear engineers and scientists have shown much less concern for safety than their counterparts in the west. This is borne out, it was claimed, by their lack of interest in building protective containment structures (which are mandatory in western countries) around Soviet reactors. It turns out that both accusations have some foundation, and both factors had their part to play in the Chernobyl accident. But other forces were also at work.

Within six months of the accident the Soviet authorities had provided to the world at large, through IAEA, a detailed account of events leading up to, during and after the explosion and radiation release (see Chapter Seven). Their frank and speedy report was welcomed internationally for the light it threw on the whole incident. It was widely seized upon by politicians and the nuclear industry who were anxious to persuade people in the west that Chernobyl was a one-off event, and similar accidents could not happen this side of the Iron Curtain.

For example, Britain's CEGB produced a publicity leaflet entitled *Nuclear Power and Radiation; The facts,* which stated categorically that "the Chernobyl accident is physically impossible in present and proposed CEGB nuclear power stations". Prime Minister Margaret Thatcher told the House of Commons soon after Chernobyl that "the record of safety and design, operation, maintenance and inspection in this country is second to none". Complacency and political rhetoric are curious bedfellows in the nuclear safety debate!

The argument that *"It can't happen here"* rests mainly on four main areas - reactor design, construction practices, containment facilities and operator training and supervision.

REACTOR DESIGN

The light-water-cooled, graphite-moderated, pressure tube RBMK-1000 reactor was the key component in the Soviet civil nuclear programme. By mid-1986 it was providing over 60% of the country's nuclear generating capacity. The basic design had evolved directly from Russia's first nuclear reactor which began work at Obninsk in 1954, and it had operated effectively and apparently safely for over 30 years. The Russians were rightly proud of it.

However, after Chernobyl they were to acknowledge, albeit somewhat reluctantly, that it has some inbuilt design faults which certainly contributed to the disaster, even if culpable human error was the immediate

147

trigger. During cross-examination at the Vienna IAEA meeting in September 1986, the Russians confirmed that the reactor design placed heavy responsibility on their operators, and in turn the operators put the reactor in an inherently unstable condition.

Reactor instability

The Soviet report acknowledged that significant violations had occurred in the operating rules for the RBMK power station immediately prior to the explosion. Apart from flagrant disabling of the reactor's control systems, the station operators - preoccupied with their unauthorized experiment (see Chapter Seven) - failed to keep the power in the reactor above the critical level for which it had been designed.

The RBMK reactor was known to be unstable and thus potentially dangerous at power levels below 700 megawatts because of inherent design limitations, and it was forbidden to operate reactors in this state. The operators could not stop the power from falling as low as 30 megawatts during their experiment before eventually getting it back up to around 200 megawatts, and only then after much effort and with little subsequent improvement. The explosion and subsequent release of radioactivity were inevitable given that critical drop in power.

During and after the accident, the spotlight fell on other nuclear reactors elsewhere with design features similar to the RBMK-1000. It soon became clear that the Russian RBMK design is different from any other design used to generate electricity anywhere else in the world. Lord Marshall, Chairman of Britain's CEGB, told the House of Lords on 20 November 1986 that "although there were gross violations by the operators (at Chernobyl), the prime cause of the accident was inherent design shortcomings". Indeed, Academician Valerie Legasov of the Soviet review group on Chernobyl has also stressed that a similar accident would not be possible at any other nuclear reactor in the world.

RBMK design shortcomings

The RBMK is water cooled, whereas all existing British nuclear power stations are gas cooled. US nuclear stations are all water cooled, but they have cores no larger than half the size of the RBMK reactor core. The RBMK is also graphite moderated, like the atomic pile used by Enrico Fermi to create the world's first nuclear chain-reaction at the University of Chicago's Stagg Field in 1942. Many US nuclear power stations are water moderated (these are the so-called Pressurized Water Reactors, or

PWRs), but to date Britain has favoured graphite moderators in its Magnox and Advanced Gas-cooled Reactors (AGRs). Britain's first PWR will be built at Sizewell in Suffolk (see Chapter Ten).

Taken together, these two key design elements (graphite moderator and water coolant) give a recipe for disaster. Graphite and water are described by nuclear physicists as "mutually energetically reactive", meaning that when they are placed together they react in a violent way (quite literally, they cause an explosion). Many western nuclear engineers insist that the key problem with the RBMK design is the assumption that two such mutually energetically reactive materials can be safely separated from each other by the welded steel tubes in the reactor core cooling system. Failure of the welding or rupture of the tubes could give rise to serious reaction problems between the graphite and steam.

But there is an additional design flaw in the RBMK, which makes it almost unique world-wide. This is the reactor's combination of positive void and power coefficients. In effect this means that if the amount of steam in the reactor core increases (for whatever reason), the power level in the reactor also increases. More water is turned to steam, so the power level rises even further. A run-away self-sustaining chain-reaction is instantly set in motion, similar to that involved in nuclear fission itself (see Figure 1.3) and capable of producing a massive increase (surge) in power within seconds. No other commercial reactor design in the world has such an in-built design weakness, which makes the RBMK rather like a time-bomb awaiting an escape of steam (from the cooling tubes) into the core or a sudden alteration of the reactor's power level.

The third main design weakness in the RBMK is its lack of automatic fast-acting shut-down systems to close down the reactor if a fault occurs or anything goes wrong. All western commercial reactor designs include at least several independent shut-down systems which cannot be disabled by the station operators (wilfully or otherwise) and which automatically shut down the reactor safely should the need arise.

British reactor designs

The prospects of an accident similar to Chernobyl ever occurring in Britain were dismissed by the country's nuclear authorities as extremely slim. The RBMK design had been studied in detail by British nuclear experts (from Babcocks Power) in 1975 and it had been rejected outright on safety grounds. Ron Campbell who led the review team expressed concern at the lack of pressurized containment vessels in the reactor, the lack of concrete shielding around the turbines, and the relatively poor overall safety standard of the RBMK compared with western reactors. Lord Marshall (CEGB Chairman) told the British Nuclear Forum after Chernobyl that "using British safety standards, the Russian design wouldn't even get past first base".

The CEGB boasted defiantly and categorically that "an accident similar to that one could *not* happen in a British nuclear power station". The CEGB currently uses two types of nuclear reactor in generating electricity in Britain - the Magnox and the AGR (Advanced Gas-cooled Reactor). Both are gas cooled, and it is physically impossible for a positive void coefficient to occur in either design. This means that no real problems are created if water and graphite come in contact, as has already happened several times in Britain. In one incident during the 1970s several thousand tonnes of seawater flooded into the core of Hunterston AGR in Scotland, and there was no explosion like Chernobyl (the main long-term problem is corrosion of pipework within the core).

The CEGB also argue that neither type of British reactor can suffer a loss-of-coolant accident, and the graphite-moderated Magnox reactors run at temperatures well below the ignition point of graphite in air (whereas the RBMK runs above that critical temperature). Furthermore, British reactor designs include tanks of inert carbon dioxide which are permanently piped into the reactor cores, allowing station operators to send gas quickly into the reactor to keep its coolant above atmospheric pressure (and thus exclude air, so minimizing the prospect of fire in the reactor).

Hanford 'N-reactor', USA

The US nuclear authorities were also quick to declare that an accident like Chernobyl could not happen in the United States, because the RBMK design is radically different from any US commercial reactor. However, it does share some similarities with a US Government reactor at Hanford Reservation near Richland, Washington. This so-called 'N-reactor' is an aging military reactor operated by the US Department of Energy. It has been used since December 1963 to produce plutonium for the US nuclear weapons programme, and since 1966 it has also produced electricity. Hanford is graphite moderated and water cooled like Chernobyl, and it also lacks proper containment structures (which are mandatory in US civilian nuclear power plants). Its core fuel channel structures are also similar to those in the RBMK.

These similarities between Hanford and Chernobyl understandably caused some anxiety in the US, and a detailed review of the Hanford installation was ordered without delay. It began on 19 May 1986. Reviews were also ordered on four other plutonium-producing reactors in the USA. These are the heavy-water-moderated reactors at Savannah River in South Carolina which lack containment facilities and are not licensed by the National Regulatory Commission (NRC).

But Hanford appears to differ from Chernobyl in two significant ways which, the Department of Energy argues, would make an accident of that sort impossible. First, it has a negative void coefficient, which means that an increase in core temperature would lead to a decrease in power (the opposite to Chernobyl's positive void coefficient). Secondly, the control rods can be dropped automatically into the reactor core extremely fast should the need arise; much faster (1.5 seconds) than in the RBMK reactor at Chernobyl (10 seconds). Loss of control of the reactor is thus much less likely at Hanford, even should a highly unlikely sequence of mishaps occur (accidentally or otherwise). There are further comforting differences between Hanford and Chernobyl, such as its independently-powered system to circulate cooling water through the graphite blocks of the reactor core, and its horizontal fuel channels which allow cooling water to be fed in from either side.

Since Chernobyl critics of the Hanford 'N-reactor' within the United States have called for its shut-down on safety grounds. The reactor has a large confinement system designed to handle major accidents involving pipe breaks, but it has no pressure-tight containment building. The review estimated that it would cost around $1.2 billion in repairs to keep the Hanford reactor operating properly beyond the year 2000, although the Department of Energy is known to prefer closing the facility by the mid-1990s.

Other commercial nuclear reactors

There is one other type of graphite-moderated nuclear reactor in use outside the Soviet Union today, and this is the high temperature gas-cooled reactor (HTGR). By the end of 1986 there was one operating in the US and one in Germany. The general belief is that although the HTGR design does not include any pressure-tight containment structure it does have adequate in-built safety systems to ensure that an accident like Chernobyl could not occur. These include using ceramic-coated particle fuel, a high heat capacity of the graphite, high temperature stability of the core and the inert nature of the helium gas coolant (which cannot react chemically with the fuel, graphite or other reactor components).

Whilst nuclear engineers in the west have retained public confidence in

151

their ability to prevent accidents like Chernobyl from happening at home, because of the sort of design differences outlined above, the issue of Soviet-built nuclear power stations abroad has been a vexed one. Because of its suitability for producing military plutonium, the RBMK reactor has not been exported or sold beyond the Soviet Union. But the Russians have been happy to export the pressurized, water-moderated, water-cooled VVER reactor (for example to Finland and Poland).

As the accident at Chernobyl was happening, Soviet nuclear engineers were helping Cuba to install a pair of nuclear reactors (believed to be of the VVER design) near the town on Cienfuegos, about 400 km south of Miami. Whilst this caused some public concern in some parts of the United States, it caused relatively little anxiety amongst US nuclear specialists who appreciated that the units were being housed in proper containment buildings, and the water moderator is much less hazardous than the RBMK graphite moderator.

REACTOR CONSTRUCTION

An interesting ingredient in the Chernobyl post-mortem is the question of construction standards and practices at nuclear installations. Before the accident, the Soviet view was very much that nuclear accidents like Three Mile Island in the USA (1979) were the almost inevitable outcome of corner-cutting by capitalist planners and engineers driven by the uncontrollable desire to maximize profits on the project. But the argument was soon to be reversed, and the Soviet authorities were soon to find themselves in the dock facing similar accusations of corner-cutting, albeit driven by different motives.

Shortly before the Chernobyl accident the Soviet press had carried reports of violations of regulations at several nuclear power stations, including Chernobyl, which put the plants at risk. Inspectors at the Rostov power station reported 136 "serious deviations" within a year. An editorial in *The Times* on 19 May 1986 records that "when they refused to pass the work, the chief construction engineers simply signed the document themselves, since they and the local Party were prepared to put plan targets before safety".

Construction deficiencies

The safety of the Chernobyl plant was questioned by some locals, even before the fateful accident. For example, the Soviet magazine *Litura-*

turna Ukrayina carried an article by local writer Lyubov Kovalevs'ka early in April 1986 which was strongly critical of what she saw as slipshod practices and unprofessional workmanship during the building of the power station. A range of "deficiencies" in the quality of construction were listed, all allegedly arising from corner-cutting expedients designed to complete the plant on schedule after long and recurrent construction delays.

Quality control in large Soviet construction projects has been criticized on various occasions, and the anxieties are not confined to nuclear power developments. But the risks introduced by deficient construction are seriously magnified in nuclear developments. The practice of *'storming'*, that is speeding up a project which has fallen behind schedule (eg by relying on large-scale overtime working to meet target completion dates) appears to be common in the Soviet Union.

Such *'storming'* is encouraged by various factors. One is the standard Soviet practice of establishing tight production targets (quarterly, annual and five-year), with workers' pay and bonuses depending on meeting them. A second factor is the country's highly bureaucratic supply system, which often produces long delays in delivery of materials and products. Another is the country's heavily overworked and much overloaded transport network which makes prompt deliveries difficult to guarantee.

The extent to which construction deficiencies contributed to the Chernobyl accident will doubtless never be fully known. Speculation and allegation will persist. But there are question marks over the overall quality of work at the plant. Even if this was a relatively minor ingredient in the accident, it is still very serious, especially given the large and growing number of Soviet nuclear power stations (including over 20 RBMKs operating or being built by mid-1986), presumably built under similar conditions and to similar standards.

Nuclear power station operators in the west insist that building standards and licensing requirements are much tougher outside the Soviet Union, and such construction deficiencies would simply not be acceptable. On this count Chernobyl *"could not happen here"*, it is argued, apparently unanimously (at least within the nuclear industry) in the west.

CONTAINMENT FACILITIES

Part of the debate over whether Chernobyl *"could happen here"* centres on the question of containing releases of radioactivity from damaged nuclear reactors, and preventing them from leaking out from the power station into the environment. The bulk of the damage caused by Chernobyl arose because of the large radiation cloud which leaked uncontrollably from the plant for ten days and spread over much of mainland Europe (see Chapter Five). Had it been possible to contain this material on site, then the accident would certainly not have been so serious and it would quite probably never have been detected outside the Soviet Union.

Western-style containment

Many nuclear engineers in the west have claimed that even if an explosion on the scale of Chernobyl was to occur in a reactor in Britain or the United States, the hazardous fission products would be trapped within the containment structure of the plant. It is mandatory in western countries to include some form of protective shell structure, capable of withstanding pre-determined pressures (ie anticipated types and sizes of explosions), in the basic design of civil nuclear plants. Containment can take various forms, but it normally involves some form of steel container or pressure vessel around the reactor core, set within a reinforced concrete radiation shield (see Figure 9.1). Most UK and US reactors also have strong containment buildings or domes around the reactor complex, as a protective outer shell or final defence against accidental leakage.

The evidence from Three Mile Island, where virtually all of the radiation leakage from the damaged reactor core was effectively trapped

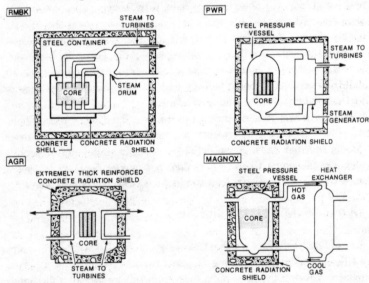

Figure 9.1   Containment and protection structures in different types of nuclear reactors

within the sealed containment building, and from other reactor accidents elsewhere, does show that properly designed and constructed containment structures can play a significant role in preventing the accidental release of radioactive material into the environment.

The Russians have traditionally shunned western-style containment structures, dismissing them as needless extravagances favoured by capitalist constructors as a means of increasing their profits. Ivan Yemilianov, deputy chief of the Soviet Energy Equipment Institute and one of the chief designers of the RBMK, insisted after Chernobyl that "experience shows that containment vessels do not guarantee complete safety. Those of the type used in the west cannot stand very high pressures thrusting upwards and are easily fractured. The vessels are also dangerous because personnel may be lulled into a false sense of security by them".

Soviet strong boxes

The Russians have favoured a *preventive* approach to nuclear safety, in the belief - proven seriously flawed by Chernobyl - that nuclear accidents can be prevented. This approach has been based on using modular 'strong boxes' as a simpler and cheaper but equally effective means of containment.

The RBMK design has three concentric 'strong boxes' nested together like a set of stacking Russian dolls, with a steel container around the reactor core inside a concrete radiation shield, all surrounded by an outer concrete shell (see Figure 9.1). The concrete container and shell were both ruptured in the two violent explosions when the accident occurred (see Chapter Three), and there was no overall containment dome or building capable of trapping the radiation which leaked out. The Chernobyl reactor core was housed in a simple multi-storey shed structure which was not designed to withstand a violent explosion, and neither did it.

Media coverage of Chernobyl strongly implied that similar accidents are extremely unlikely to occur in US and British nuclear plants because they all have adequate containment facilities. Not everyone agrees. Critics of the US nuclear energy programme point to the fact that nearly half of the existing US nuclear plants do not have proper containment domes. Definition of what is 'proper' and 'adequate' in this context is extremely difficult, but it is a fact that 39 US nuclear plants (all designed by General Electric) have a 'pressure suppression' type of containment similar to Chernobyl before the accident. All 39 are believed to be susceptible to a Chernobyl-type accident, involving potentially massive releases of radioactivity to the environment.

Since Chernobyl there are signs that the Russian authorities might be prepared to review their approach to safety, possibly giving greater emphasis to proper containment facilities. This does not necessarily mean that containment domes will be built at existing Soviet nuclear power stations, or even included in the design of stations currently being built or planned. But their blanket insistence that such facilities are unnecessary has at least fallen into serious questioning, at home and abroad!

OPERATOR PRACTICES

Western nuclear experts have stressed that the RBMK design placed too much responsibility on the shoulders of the reactor's operators. The Russians also conceded this at Vienna (September 1986), when questioned about the unauthorized experiment and the disabling of the reactor's safety systems.

The Chernobyl technicians had much greater scope than their counterparts in the west to interfere with the running of the reactor. They could, for example, and apparently with ease, switch off all of the station's safety systems in preparation for the experiment and deliberately withdraw most of the control rods from the reactor (see Chapter Seven). These were critical ingredients in the accident, and once they had been established the reactor inevitably continued to surge with power (because of its positive void coefficient). The operators could only try in vain to bring it back under their control.

Automatic shut-down facilities

Had there been an automatic shut-down system in the Chernobyl reactor, which could not have been disabled by the operators, the critical conditions for the accident would probably never have been allowed to arise. But, unlike all commercial nuclear reactors in use in the west, the RBMK reactor was not designed to cope automatically with operator error or with accidental failure.

A similar accident could not happen in Britain, the CEGB claim, because UK nuclear safety philosophy and practice are founded on safety in depth. This emphasizes in-built safety systems which can shut down the reactor automatically, quickly and safely if its stability and safety are threatened. These safety systems are incorporated in the reactor design from the very first stages, and they cannot be disengaged or over-ridden by operators. UK nuclear plants have in-built surveillance systems which rapidly detect anything abnormal within the reactor and sound appropriate alarms. They also activate in-built emergency multiple automatic shut-down systems (mechanical and electrical) which are designed to be fail-safe. It is claimed to be impossible for operators to deliberately (or accidentally) short-circuit these in-built safety systems; back-up systems would automatically shut down the reactor if they were interfered with.

These automatic safety systems in British power stations are complemented by strict security checks. For instance, operators are required to get written permission to gain access to the key which operates the overall reactor safety system, and warning signs appear in the station's control room if an operator attempts to disable any of the safety systems (and the back-up safety systems are automatically activated).

Personnel

The UK civil nuclear industry is overseen by a well trained and highly professional inspectorate, the Nuclear Installations Inspectorate (NII), an independent branch of the Government's Health and Safety Executive. The NII is the country's nuclear watchdog, with power to inspect any nuclear power station at any time, and to close down any such station or restrict its operations if it regards this as necessary.

Very tight control is kept over tests or any changes in normal operating procedures in British nuclear power stations. If any such changes are required, detailed written plans must be submitted to a group of expert engineers drawn from four separate departments within the CEGB or to the power station's nuclear safety committee. Approval must also be got from the NII.

UK nuclear plant operators are highly trained graduates who have worked in the CEGB for at least several years and undergo at least a year's strict training devoted specifically to nuclear reactors. Their training programme is both theoretical and practical, and it includes the use of simulators to allow the trainees to rehearse routine and emergency practices under simulated conditions. The training of Russian technicians and operators appears to have been particularly deficient in this area. The Soviet report to IAEA admitted that their training programme should in future be improved specifically to deal with unusual situations (most probably using western-style simulators), to avoid the unknown or indifferent type of human reaction which seems to have occurred at Chernobyl.

CONCLUSIONS

Shirley Williams, President of Britain's Social Democratic Party, said in *Today* newspaper soon after Chernobyl that "the genie of nuclear power cannot now be pushed back into the bottle from which it escaped in 1945". She voiced the fears of many ordinary folk who were alarmed by the accident and anxious to find out if a similar one *could* happen closer to home.

The person in the street must rely entirely on the statements offered by nuclear engineers and scientists. These have, on the whole, been very reassuring. The nuclear industry in the west has almost spoken with one voice on the matter, and concluded that similar accidents could not happen in Britain and the United States. We have seen above the main threads of their defence - western nuclear reactors are inherently safer in design, better built, more effectively contained and run by more highly trained professionals than Soviet reactors (especially the RBMK design).

But the nuclear industry may not be the proper source of reliable comfort, because it is far from impartial. It clearly has a strong vested interest in convincing people that nuclear power is safe and reliable. To even concede that Chernobyl might happen here would severely damage public confidence in nuclear power, which has on the whole been fragile since the accident in any case (see Chapter Ten). Lord Marshall of Britain's CEGB is typical of the confident nuclear leaders in the west. He told the House of Lords on 20 November 1986 "we are satisfied that there is no narrow technical issue which we in the west could or should learn from the Chernobyl disaster. We have very well established safety rules which we follow meticulously. The Russians have chosen to ignore many of them, and sadly they have now paid the price".

The nuclear industry has a responsibility to put the public mind at rest. But it also has a responsibility to reveal the truth, and some critics of nuclear power claim that there is little justification to the industry's monolithic view that *"it can't happen here"* . The kind of defence offered by the industry and outlined in this chapter is doubtless well intended and, insofar as ordinary folk are in a position to judge, logical and (presumably) factually correct.

But, critics argue, the industry may be asking, and in turn answering, the wrong question! Public confidence may have been won back after Chernobyl by assuring people that the exact sequence of events involved in that accident - including the running of the experiment, disabling of safety systems, withdrawal of control rods, problems of running a reactor at low power, failure to keep control of the reactor, explosion and release of radioactivity - *"can't happen here"* . But the industry has not put people's minds at rest that an accident of similar size or seriousness, perhaps caused by a completely different set of factors and in a completely different way, *"can't happen here"* .

In fact the industry has to some extent drawn attention to the prospect that such an accident is at least possible (no matter how unlikely it might be in purely statistical terms) by remaining so silent on the issue. The anti-nuclear lobby and the public at large would have been far more comforted after Chernobyl had the industry been able to say with confidence that a nuclear accident which caused similar levels of radioactivity to leak into the environment *"can't happen here"* . So much the better if it were possible to openly demonstrate the basis for that confidence!

Some critics of Britain's Magnox and AGR reactors believe that such accidents are possible, perhaps through failure of the cooling systems (for example by rupture of coolant tubes or pump failure) which might lead to loss of control of the reactor followed by a violent explosion (like Chernobyl) and subsequent release of dangerous fission products. Whether UK-style containment structures are capable of withstanding such explosions and trapping the radioactivity is open to question, they insist. Similar fears have been voiced by the anti-nuclear movement in the United States.

So the overall conclusion must be that Chernobyl itself probably *"can't happen here"* in the west, but that does *not* remove the possibility of an equally (or more) serious accident happening for different reasons. Many people inside the nuclear industry and beyond had hoped that Three Mile Island would be the last serious hiccup in nuclear power's march forward to become the best, cheapest and most reliable source of energy in the next century. But a technology so firmly founded on taming such inherently unstable reactions must always be liable to fail on occasions. The real questions are *where, when* and *how big* will the next failure be? Only time will tell.

10
PUBLIC CONFIDENCE IN
NUCLEAR POWER?

Chernobyl cast a long shadow over public confidence in nuclear power. Peter Bunyard had concluded after the accident at Three Mile Island in 1979 that "nuclear power has lost its innocence - no more does the public implicitly believe officials who tell it that nuclear power is cheap, clean and safe". Britain's House of Commons committee of inquiry on nuclear power concluded in the summer of 1986 that one lesson from Chernobyl is that "public opinion will play a much larger part in deciding the future of nuclear power than is usual with questions of science and technology". We look in this final chapter at how public opinion over nuclear issues was affected by the accident at Chernobyl, and how Governments responded to these swings in the light of their own commitment to nuclear safety and strategies.

Faustian bargain

At the end of 1985 there were a total of 374 nuclear reactors operating in 26 countries and a further 157 reactors under construction (Table 10.1). Nuclear power was producing 15% of the world's electricity in 1985, compared with a mere 8% in 1980 and an anticipated 20% by 1990. It appears to be here to stay, promising reliable supplies of energy in the future when conventional fossil fuel power sources (coal, oil and gas) start to dwindle.

But nuclear power has always had a mixed reception, and some describe it as a *'Faustian bargain'*, or a bargain with the devil. Faust, in German legend, was a magician and alchemist who sold his soul to the devil in exchange for power and knowledge. The relevance of this *'Faustian bargain'* to Chernobyl is more than passing. One of the late C. S. Lewis's more popular books was called *Screwtape Letters*, and it takes the form of a series of letters from the devil (*Screwtape*) to his nephew *Wormwood*, advising him how to spoil the otherwise perfect

Table 10.1 *Status of nuclear power by country at end of 1985*

country	in operation units	in operation total MWe	under construction units	under construction total MWe
Argentina	2	935	1	692
Belgium	8	5,486	0	
Brazil	1	626	1	1,245
Bulgaria	4	1,632	2	1,906
Canada	16	9,776	6	4,789
China	0		1	300
Cuba	0		1	816
Czechoslovakia	5	1,980	11	6,284
Finland	4	2,310	0	
France	43	37,533	20	25,017
German Dem. Republic	5	1,694	6	3,432
Germany, Fed.Republic	19	16,413	6	6,585
Hungary	2	825	2	820
India	6	1,140	4	880
Iran	0		2	2,400
Italy	3	1,273	3	1,999
Japan	33	23,665	11	9,773
Korea, Republic of	4	2,720	5	4,692
Mexico	0		2	1,308
Netherlands	2	508	0	
Pakistan	1	125	0	
Philippines	0		1	620
Poland	0		2	880
Romania	0		3	1,980
South Africa	2	1,840	0	
Spain	8	5,577	2	1,920
Sweden	12	9,455	0	
Switzerland	5	2,882	0	
United Kingdom	38	10,120	4	2,530
USA	93	77,804	26	29,258
USSR	51	27,756	34	31,816
Yugoslavia	1	632	0	
WORLD TOTAL	374	249,625	157	141,942

SOURCE: *IAEA Bulletin* (Summer 1986), page 67

world (which is, after all, the devil's self-professed purpose in life!).

By sad paradox, the name *Chernobyl* means *Wormwood* in Ukrainian. Some have speculated that the nuclear accident itself had been foretold in Biblical prophecy in *Revelation* (8:10-1), which describes how "a great star, blazing like a torch fell from the sky ... the name of the star is Wormwood (ie bitterness) ... and many people died". The fickle circle of fate might have been made curiously complete by the accident at Chernobyl!

Loss of faith

Chernobyl was a serious blow to public confidence in the nuclear industry around the world. In the wake of the accident there were many protests against nuclear power, and public opposition was galvanized in a cohesion rarely seen before. Governments were forced to think very carefully about their future energy plans, and to reflect soberly on what role nuclear energy would play in them. The nuclear industry also had to take a close look at itself, to bolster its flagging credibility and massage its severely dinted ego.

To many people around the world the Chernobyl accident was the last straw. Through the 1980s opposition has been rising to most phases of the nuclear fuel cycle (Figure 10.1), not simply the nuclear power stations themselves. There were many issues of safety involved, including the siting and operation of plants for fuel preparation (enrichment and conversion) and for reprocessing used fuel from fission reactors, the transport of radioactive materials from site to site, and the storage and disposal of waste products.

The period immediately after Chernobyl was critical for the whole future of nuclear power around the world, given the real likelihood that public faith in nuclear energy would fall sharply. But just how serious and long-lasting was this loss of faith? The early speculations were that public confidence, long shaky, must now crumble completely. This turns out not to have happened.

REACTIONS IN THE SOVIET UNION

The accident at Chernobyl doubtless created anxieties and caused problems in the Soviet Union. But overall it appears to have done little to deflect the country from the path mapped out by Lenin at the time of the Great October Socialist Revolution in 1917. He declared that "Communism is Soviet power plus the electrification of the whole country", seeing electrification as the technical basis for socialist construction. Tom Wilkie and Roger Milne, writing about Chernobyl in *New Scientist* soon afterwards, concluded that it "will be a crushing blow to both the military and the civil nuclear power programmes in the USSR". There are no signs

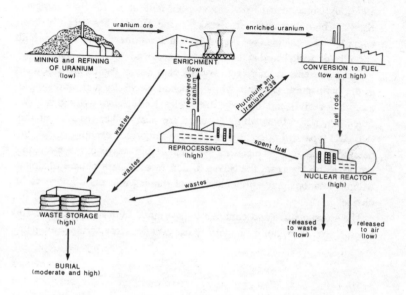

Figure 10.1 The nuclear fuel cycle

that this has been the case.

No formal demonstrations against nuclear power within the Soviet Union have been reported in the west before or since Chernobyl. But the KGB had to move quickly to prevent an *ad hoc* coalition of academics and dissidents (the so-called *'Trust Group for Peace'*) from gathering outside Moscow's Vachtangova Theatre on Tuesday 20 May 1986 to collect signatures for an anti-nuclear petition to hand in to the Kremlin.

Faith in the RBMK

At the time of the accident Russia was getting 11% of its electricity (Figure 10.2) from nuclear sources. There were 51 nuclear reactors (Table 10.1) of two types - the RBMK and the VVER. The VVER is a Russian designed and built water-cooled, water-moderated reactor. It is similar to commercial PWRs built in the west, like Three Mile Island in the USA and the planned new development at Sizewell in England. The VVER has been sold overseas, including 15 to eastern Europe and 2 to

Finland.

RBMKs like Chernobyl provide the bulk (60%) of Soviet nuclear power. The first full-scale RBMK was completed near Leningrad in 1973, and since then 15 more have been built at 5 sites (in Leningrad, Kursk, Chernobyl and Smolensk). One RBMK has recently been upgraded at Ignalina (see Figure 1.1), and a further 15 were being built or on the drawing board at the time of the accident.

This represents a massive investment of Soviet capital and resources in the multi-purpose plants which generate electricity and produce plutonium for civil and military use. Peter Kelly, also writing in *New Scientist*, noted that "although there are many other reactors of this (RBMK) type still operating, the Soviet authorities cannot exclude the possibility that there is a generic design fault in this type of reactor. If this proves to be the case, the Soviet Union faces a stark choice of doing without weapons grade plutonium, or of running reactors it knows to be unsafe".

The Russians had and continue to have great faith in the RBMK-1000 reactor. But its reliability inevitably came under close scrutiny after the

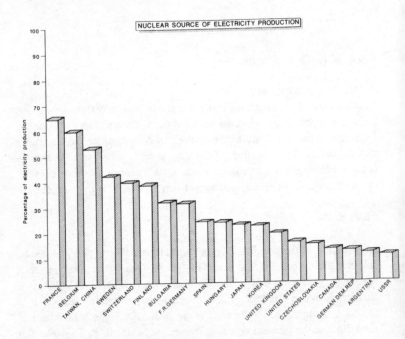

Figure 10.2 *Nuclear share of electricity production by country, 1985.*
After Bennett & Skjoeldebrand (1986)

accident, and the remaining 27 were shut down temporarily for safety checks. Adaptations were added to make it impossible for operators to over-ride the RBMK's safety systems, and they were also equipped with safeguards to ensure that the control rods are inserted at all times. But the Russians desperately needed the power generating capacity of the RBMKs, and they were soon brought back on line. Chernobyl's four reactors were providing 15% of the total Soviet nuclear power capacity at the time of the accident, and the three undamaged ones were restarted within six months.

The Soviet authorities recognized the need to put their own nuclear house in order in the wake of Chernobyl, and - just as important - to be seen to be doing so by the eyes of the world. It was clearly necessary to convince their own people, as well as the rest of the world, that the RBMK reactors can safely be allowed to continue in operation. Just about all responsibility for the accident had been placed on the shoulders of the Chernobyl workers (see Chapter Seven). The Politburo exonerated the RBMK reactor design, although a strong hint of official concern over the RBMK came with the dismissal from office in July 1986 of Ivan Emelyanov, head of the design team that had produced the RBMK in the 1960s.

The authorities also recognized the need for much better management and control of the Soviet nuclear industry. A new ministry was set up with overall responsibility for the design and construction of new reactors, to "raise the level of management and responsibility in the nuclear power industry". A significant boost of confidence in the RBMK was given by the appointment of Nikolai Lukonin (former Director of the RBMK stations at Leningrad and Ignalina) as Minister in this new All Union Ministry of Nuclear Energy. The Politburo also asked all ministries and government departments to introduce new measures to ensure the safety of existing Soviet nuclear reactors.

Nuclear expansion

Soviet Premier Mikhail Gorbachev noted on 14 May 1986 that "the indisputable lesson of Chernobyl to us is that in conditions of the further development of the scientific and technical revolution, the questions of reliability and safety of equipment, the questions of discipline, order and organization assume priority importance". This implied a continued commitment to nuclear power after Chernobyl which was echoed by Mr A. Petrosyants, Chairman of the Soviet State Committee on the Utilization of Atomic Energy, who declared that "the path of nuclear power development and the growth of nuclear power in the USSR are to remain unchanged".

Russia is still committed to an energy policy based on accelerated growth of nuclear power. Energy in general, and nuclear power in particular, are key elements in Gorbachev's drive to double the size of the Soviet economy by the year 2000. The Kremlin's Five Year Plan for 1986-90 has a target electricity output of 390,000 thousand megawatts by 1990 (compared with an actual output of 167,000 thousand megawatts in 1985), and a planned nuclear contribution of 20% (compared with 2% in 1979 and 11% in 1986).

Russia has one of most active nuclear construction programmes in the world. Various factors have encouraged this, including the country's lack of anti-nuclear protests and its well developed nuclear construction industry. The shortage of alternative viable and sustainable energy resources, particularly given the poor performance and weak prospects of the Soviet oil industry, has been a further catalyst. Nine new nuclear plants (VVERs and RBMKs) were being built in May 1986, and capacity was being increased at 9 existing sites. A further 30 new reactors are scheduled for construction over the next decade.

Soviet confidence in the safety of their nuclear plants is reflected in their policy (embodied in the Five Year Plan for 1986-90) of siting some reactors close to major centres of population. This means that, as well as electricity, they can also provide thermal power for city district heating schemes. The plans envisage a series of urban 'atomic-thermal power stations', each surrounded by a 'green belt' between 1 and 3 km wide which would be used for recreation and allotments. Such schemes are planned for Gor'kii, Voronezh, Odessa, Minsk and Karkhov.

The Five Year Plan also includes provision for more large nuclear power stations to be built in the western, European part of the Soviet Union (where the bulk of the country's population are concentrated) and in the Urals. Europeans might have legitimate ground for concern about this policy, especially if it includes more RBMK reactors and increases the prospects of another Chernobyl happening close to the Soviet border.

The RBMK design figures prominently in the Soviet plans to expand nuclear power. The basic unit to date has been the 1,000 megawatt reactor (RBMK-1000), as installed at Chernobyl. But a 1,500 megawatt RBMK (using two 750 megawatt turbines) has been built at Ignalina and a second is planned for Kostroma on the Volga.

The Russians are actively engaged in an atomic fusion research programme, using their own Tokamak breeder reactor design as a prototype. The hope is to use controlled fusion commercially in a fast-breeder reactor by the turn of the century, and to replace progressively the thermal RBMK and VVER reactors with breeders. They are investing heavily in fast-breeder reactors. Work started in 1973 on building a BN-350 at Shevchenko on the Caspian Sea, and a BN-600 went into operation at Beloyarsk (in the Urals) in 1980. They have even more ambitious plans for larger (800 megawatt) fast-breeder stations.

Since May 1986 the Russians have outlined ambitious plans to further increase nuclear power production, which require the continued operation of existing RBMK reactors, and construction of new ones. Chernobyl was a serious blow to their nuclear industry at the time, but it is *not* likely to substantially alter Soviet nuclear plans, designs or intentions.

REACTIONS IN OTHER EASTERN BLOC COUNTRIES

Inevitably ideological allegiance to their Soviet neighbours tempered feelings about Chernobyl in the eastern bloc countries.

There were few signs of public anxiety in Czechoslovakia, and the country has retained largely unshaken its firm commitment to nuclear power. When Chernobyl happened its five nuclear plants (Table 10.1) were providing around 15% of the country's electricity (Figure 10.2), and eleven more plants were under construction. None were cancelled, but there are signs of increased concern over matters of nuclear safety. Jiri Beranek, Chief Inspector of the country's Atomic Commission, called for a re-evaluation of all nuclear safety measures (at home and abroad) after the accident, and a tightening up of safety procedures at the five plants was expected.

Chernobyl had a more mixed reception in Poland, which was directly in the path of the radiation cloud and received dangerously high levels of fallout (see Chapter Four). There were three nuclear plants under construction at the time (Table 10.1), and the accident fuelled some anxieties over nuclear safety. Feelings were aroused particularly in Bialystok in the north-east, which received high levels of fallout from the accident (see Figure 5.4). Three hundred local residents signed a petition calling for a halt to construction of the country's first nuclear plant at Zarnowiec. But the Polish Government insisted that nuclear power is essential for the country's future. To abandon it would be folly, they argued, given a lack of viable alternatives.

Reactions to Chernobyl were mooted in Hungary. The country presently has two nuclear stations and is building two more (Table 10.1). Nuclear power provides around 23% of their electricity (Figure 10.2). The Government is committed to a marked expansion of energy resources, and the favoured solution includes a controversial hydro-electric power (HEP) scheme at Gabeikovo-Nagymaros. Hungarians were compromised. If they protested too strongly against nuclear power and condemned Chernobyl, they would increase the prospect of further expansion of HEP schemes. They chose to remain silent in the wake of Chernobyl.

The debate was more vocal in Yugoslavia. Environmentalists staged a protest rally on the first Saturday after Chernobyl (4 May) urging their Government to seek compensation from the Soviet Union for any damage the accident caused. The Government was already facing opposition to its sole nuclear power station (Table 10.1) before Chernobyl from a fairly strong anti-nuclear lobby, and it altered its energy strategy after the accident. It announced its intentions (on 12 May 1986) to postpone any decision on building a second nuclear power plant at Prevlaka near Zagreb, pending the outcome of a long-term review of the country's energy prospects. Instead of expanding nuclear power, it decided to concentrate on energy conservation schemes and the development of HEP resources. A second wave of cancellations came on 21 May 1986, when the Government announced its decision to cancel plans to build a further nuclear plant in the country's main wheat-growing area.

Bulgaria had much more to lose from a loss of public faith in nuclear power. In 1985 the country was getting around a third of its electricity (Figure 10.2) from four VVER reactors, with a combined capacity of 1,632 megawatts (Table 10.1). It was clearly committed to nuclear expansion, aiming to provide 40% of its power from nuclear plants by the year 2000. A fifth VVER was expected to be ready by the end of 1986 and a sixth was being built at the time of the accident (both with Russian technical assistance). The Bulgarian authorities had no change of heart or intentions over their nuclear programme after Chernobyl.

REACTIONS IN WESTERN EUROPE

Chernobyl came at a bad time in European nuclear history. The preceding months had witnessed scattered opposition to nuclear power across Europe, and there was continued and growing public resistance to nuclear developments. Reactions to the accident varied considerably from country to country.

Scandinavia

Finland relies heavily on nuclear power. It has two nuclear stations (each with two reactors), one at Lovissa on the south-east coast and the other at Olkikvoto on the south-west coast. Between them they were producing around 30% of Finland's electricity in 1985 (Figure 10.2). But there is also strong public opposition to nuclear power, which was a hotly contested issue within the country's four party centre-left coalition in May 1986. The clear damage which Chernobyl fallout was causing to the Lapp herdsmen who roam across northen Scandinavia (see Chapter Eight) was a particularly sore point. After Chernobyl the Government shelved plans for a fifth 1,000 megawatt nuclear plant, at least until after national elections due in March 1987.

Denmark has no nuclear power stations (Table 10.1), and no commitment to a nuclear programme. Although the Government had been exploring the prospects of nuclear power it decided after Chernobyl not to press ahead in a nuclear direction. Nuclear power was clearly too hot an issue (literally as well as figuratively) and too risky an investment; the time was obviously not right to press ahead at this stage.

Sweden first detected the radiation cloud from Chernobyl and tipped off the rest of the world (see Chapter One). Energy Minister Birgitta Dahl and her Government could scarcely contain their anger over the Soviet accident. Yet Sweden had a heavy investment in 12 nuclear plants (Table 10.1) at the time and depended on them for 40% of the country's electricity (Figure 10.2). But even before Chernobyl the barometer of public confidence had swung away from nuclear power, and the Government was committed to honour the outcome of a 1980 national referendum. This allowed no further nuclear reactors to be built, and required the shutting down and eventual dismantling of the 12 existing ones when they reach the end of their useful working lives. Shut-down was scheduled to start in 1995 and should be completed by 2010. The main impact of Chernobyl on public attitudes in Sweden was to lend weight to the lobby calling for that deadline date to be brought forward. In mid-May 1986 the Government announced that it may indeed do just that.

Central and southern Europe

France is the world leader in civil nuclear investment. The state electricity utility (Electricité de France) relies on its 43 reactors (Table 10.1) to produce a world record-breaking 65% of its electricity (Figure 10.2). Most (38) are PWRs, but they also have 4 natural uranium gas-cooled graphite-moderated reactors and a fast-breeder. France is also the most pro-nuclear country in Europe, with relatively little anti-nuclear feeling and few protests or public demonstrations against nuclear power. There had been some public opposition over the country's first Super-Pheonix breeder reactor at Malville (which had gone on-stream in January 1986) and over the siting of radioactive waste dumps in France.

Chernobyl seemed to wake up latent anti-nuclear anxieties in France, and serve as a focus of some public protests, even though the Government insisted that fallout from the accident was not being deposited over French soil (see Chapter Five). Around 5,000 people marched through Paris on Saturday 24 May 1986 in an anti-nuclear protest with the theme 'Chernobyl; never again'. On other occasions large crowds gathered around the perimeter fences of some French nuclear power stations to protest against their Government's nuclear policies. Overall France appears to have weathered the Chernobyl storm without serious problems or set-backs to its nuclear programme. Nineteen new PWR plants were under construction at the time, and confidence in the country's nuclear power programme was running so high after Chernobyl that the government announced plans to expand it even further. The new Government elected in May 1986 was known to be not entirely comfortable with the nuclear policies it had inherited from its predecessors, but it made no major alterations to existing nuclear policies.

Switzerland had five nuclear reactor running in 1986, with a total capacity of 2,882 megawatts (Table 10.1). This is around 40% of the country's total electricity requirements (Figure 10.2). The Government had confidence in nuclear power and in its own nuclear safety procedures. But the Swiss Ecology Party called on them to phase out the five existing reactors by the turn of the century, and to cancel construction of a sixth plant at Kaiseraugst, near Basle. The party also campaigned for a national referendum on nuclear power after Chernobyl. It achieved success in none of these three areas.

The Austrian Government decided in mid-May 1986 to dismantle the country's only nuclear power station, which had been completed several years earlier and was at the time awaiting a licence to press ahead into full operation.

Italy had three nuclear power stations operational at the time of the accident (Table 10.1), providing less than 10% of the country's electricity. Three more plants were under construction, which would more than double their nuclear capacity (Table 10.1). Chernobyl triggered a storm of public opposition to nuclear power. On Saturday 4 May an estimated 100,000 anti-nuclear protestors marched through Rome, led by leaders of the Italian branch of the World Wildlife Fund and the Environment League, who placed a wreath at the Soviet Embassy bearing the pointed message "For the present and future victims of Chernobyl". That same weekend a poll was taken of how Italians view nuclear power, and published in the Rome daily *La Republica*. It found that 79% of Italians thought no new nuclear plants should be built in Italy, and 33% thought the three existing ones should be closed immediately. Only 52% of the people interviewed said they believed government figures on radiation levels.

Chernobyl came at a bad time politically in the Netherlands, in the run up to a national election. It threw up nuclear safety as a major election issue and gave a golden opportunity for opposition parties to win public support by making anti-nuclear statements. In May 1986 the country had two nuclear plants, with a combined capacity of 508 megawatts (Table 10.1), less than a tenth of its electricity needs. Before Chernobyl the Dutch Government had been drawing up plans to build two more nuclear plants. Conscious of the strong public opposition to these plans, and the impact this might have on people's voting preferences, the Government neatly side-stepped the issue of nuclear safety and shelved them. It also announced its intention to carry out a special safety study of the two existing nuclear stations.

West Germany and Great Britain were also preparing for general elections when the accident happened, and inevitably nuclear issues were given a much higher profile than usual by all interested parties.

REACTIONS IN WEST GERMANY

The shock-waves of Chernobyl were felt perhaps most acutely in West Germany, a country with a strong anti-nuclear lobby, an active Green Party and a heavy reliance on nuclear power. Its 19 nuclear plants produce 16,413 megawatts of electricity (Table 10.1), around 36% of the country's total (Figure 10.2). A further six plants were due to be commissioned by 1990, bringing the nuclear capacity to around 50%.

Nuclear protests

Anti-nuclear feelings in West Germany were already high before the accident. The main focus of public opposition had been the site of the proposed nuclear waste reprocessing plant in a forest clearing near Wackersdorf in Bavaria (which, ironically, was to receive among the country's heaviest fallout of radioactivity from Chernobyl). This DM

171

6,000 million plant is designed to produce fast-breeder plutonium fuel and handle spent fuel elements (see Figure 10.1) from the country's light water reactors. In December 1985 and January 1986 there had been several clashes between police and between 7,000 and 8,000 protestors occupying the site.

Things came to a head three weeks after Chernobyl, when an estimated 20,000 demonstrators converged on Wackersdorf for an anti-nuclear rally. It was a violent affair, with two days of bitter fighting during which demonstrators threw stones, steel balls and even petrol bombs, and tried to cut or bulldoze their way through the site fences. Some 400 people were injured, including 62 police, and the crowd finally dispersed only after tear-gas had been sprayed on them from police helicopters.

Other West Germans wasted no time in voicing their anger and concern about Chernobyl. During the first weekend after the accident (4-5 May 1986) around 15,000 demonstrators stopped traffic for an hour in Munich. Other large anti-nuclear demonstrations were held in Dusseldorf and Hesse, and some 6,000 protestors assembled on the site of a planned nuclear waste dump at Gorbelen in Lower Saxony and had to be dispersed by police with tear-gas.

Chernobyl fuelled West German anxiety over nuclear power. Psychoanalysts diagnosed many cases of what they described as *'Atom Angst'* (*nuclear fear*). A poll carried out by the Emnid Research Institute in Bonn and published in *Der Spiegel* on 12 May 1986 found 69% of West Germans opposed to any expansion of the country's nuclear energy programme, compared with 46% in March 1982. Public opposition was to peak at 83% by the end of May. These pronounced swings in public attitude came at an inconvenient time for the Government, preparing for national elections in eight months time.

Nuclear politics

All contending parties realized the seriousness of Chernobyl, and the implications of the accident became a key ingredient in the political debate. The minority Green Party (Die Grunen) and the Social Democrats (SPD) - the main opposition party - demanded a steady abandonment of nuclear power in West Germany. The Free Democrats (FDP), the minority party in the coalition Government, called for a review of the country's nuclear reprocessing industry, with a special emphasis on Wackersdorf. Chancellor Kohl's Christian Democrats (CDU) insisted that an accident like Chernobyl could not happen in West Germany because of the country's more advanced nuclear technology.

It quickly became apparent that the Social Democrats' anti-nuclear stance could win them widespread public support, perhaps enough even to let them win the election outright. The Green Party saw an opportunity to strengthen its position, and made its political co-operation with the SPD conditional on their support of the call for immediate closure of all nuclear plants in West Germany. Chancellor Kohl tried to dampen the anti-nuclear argument by pointing out early in June 1986 that simply halting the country's own nuclear programme would not necessarily make West Germany safe from nuclear accidents, because there would still be 66 nuclear plants in surrounding countries. Public opinion polls in mid-May 1986 put combined support for the SPD and Greens ahead of the CDU and FDP.

It was widely speculated that the first victim of the marked switch in public attitudes away from nuclear energy might be the experimental fast-breeder reactor at Kalkar on the Rhine, which West Germany had built in partnership with the Dutch and Belgians. Building work on the site had been completed by May 1986 but it did not have a full operating licence from the state government. The problem could turn out to be very serious, it was felt, because without Kalkar the need for Wackersdorf becomes less defensible. There were also fears that nuclear plants under construction in Baden-Wurtemberg, Schleswig-Holstein and Bavaria may be threatened by the renewed anti-nuclear feelings.

It turns out that none of these potential casualties became real victims (at least over the next 18 months), and the German authorities appear to have largely contained public anxieties over nuclear safety.

REACTIONS IN BRITAIN

Chernobyl also came at an awkward time for Britain's pro-nuclear Conservative Government. By May 1986 the UK had 38 nuclear reactors in use, with a combined output of over 10,000 megawatts (Table 10.1) or just over 20% of the total electricity required (Figure 10.2). Four more reactors with a total capacity of 2,530 megawatts were being built. UK demand for electricity was growing at a rate of around 4% a year, and Energy Secretary Peter Walker had launched ambitious plans to expand the country's nuclear programme. Top priority was the construction of a new generation of US-style PWRs, the first planned for Sizewell in Suffolk.

The national election was due in June 1987, and nuclear power was clearly going to be a tricky area for Margaret Thatcher's Government to deal with. The Prime Minister insisted in the House of Commons on 8 May 1986 that "the record of the nuclear industry in this country has been excellent because in civil installations there have been no fatalities in this country ... Provided we go on, as we shall, in safety, in design, manufacture, mode of operation and maintenance, I believe that record will continue and give the public confidence once again".

But this confident view was not shared by everyone in Britain, and the opposition parties (especially Labour) wasted few opportunities to exploit the electorate's doubts about nuclear safety. There were three main areas of public concern and political debate: disposal of radioactive wastes, reprocessing of spent fuel and development of the PWRs.

Radioactive wastes

Most parts of the nuclear fuel cycle generate waste products (see Figure 10.1), and many of these are radioactive. Some - the so-called *'low level wastes'* - pose little direct danger to human health. These are mainly items worn or handled by people in nuclear laboratories (such as rubber gloves, clothing and syringes), and they account for about 85% of Britain's nuclear waste. Low level wastes can be safely stored with acceptable risk. But other wastes have dangerously high levels of radioactivity which, for some radionuclides, will persist for vast stretches of time (see Table 2.3). Safe storage of *'intermediate level wastes'* (such as filters, sludges and resins) and *'high level wastes'* (such as used fuel) poses serious problems, and is one of the most contentious areas of nuclear safety.

Throughout 1986 the Government had faced stiff opposition and a barrage of public protests against its plans to bury low and intermediate level nuclear waste underground at geologically stable onshore sites. NIREX (the Nuclear Industry Radioactive Waste Executive), the public agency responsible for radioactive waste disposal in Britain, had explored all the options and identified four sites in England suitable as waste dumps. These were Bradwell in Essex, Elstow in Bedfordshire, Fulbeck in Lincolnshire and South Killingholme in South Humberside. Once these potential dump sites had been named, local communities were incensed and mobilized considerable opposition (including their local Conservative MPs). Protests against the NIREX plans - including occupation of the dump sites, road-blocks against NIREX vehicles and petition writing - were well organized and received much publicity.

The timing of the NIREX announcement (in the run-up to the election) was fortunate for the protestors, because the Thatcher Government needed to dampen down anti-nuclear feelings around the country as quickly and effectively as possible in order to minimize loss of electoral support. On 2 May 1987 the Government announced, in its reply to the Commons Select Committee on the Environment report on nuclear waste, that it will only allow low level radioactive waste to be buried at the dumps.

This was seen as a major climb-down for the Government and a major victory for the anti-nuclear lobby. The Government clearly needed to save face as much as possible. It insisted there were no scientific grounds for restricting dumping to low level waste, but stressed that "this is an area where it has proved particularly difficult to bridge the gap between scientists' assessment of risk and the honestly-held perceptions of the local communities". Intermediate wastes will now be stored pending development of a new deep disposal site or until radioactivity has reduced enough for disposal.

Reprocessing

Over 90% of the radioactive waste produced from civil power generation over 30 years in Britain (including spent fuel from the 11 Magnox power stations of CEGB and South of Scotland Electricity Board) is presently being stored at British Nuclear Fuel Ltd's (BNFL) nuclear fuel reprocessing plant at Sellafield in Cumbria (see Figure 1.1). Some of it is in liquid form, thus it is dispersable and a potential risk to people and the environment.

The biggest worry to many people is the safety of the plant for reprocessing Magnox fuel, which is over 20 years old and - according to some critics - badly maintained. Public confidence in Sellafield has been rocked by a series of accidental leaks and releases from the site, especially since 1979 (when better records have been made available for public scrutiny). During 1986 there were several serious leaks of radioactive waste directly into the Irish Sea, episodes of contamination on site and a fire at the waste disposal site. The media were quick to pick up on these, and public anxieties (locally and nationally) rose accordingly.

The Government's Health and Safety Executive (HSE) gave BNFL an ultimatum early in December 1986 - make changes at plant or it will be closed. The HSE requested improvements in management, control and instrumentation, containment, ventilation, decontamination, transfer of radioactive materials, maintenance and staff training. Mr John Rimington, Director General of the HSE, said the report was designed as a "major jolt to the management" at Sellafield. BNFL, conscious of their own poor track record and equally aware of the need to court public support, welcomed the audit as "a helpful contribution to the further improvement of safety at Sellafield".

Sizewell PWR

There were fears that Chernobyl could force long delays in getting approval for the Government's controversial PWR proposals for Sizewell. The CEGB case for Sizewell and critics cases against it were debated at length and in some depth during a costly 340 day Public Inquiry led by Inspector Sir Frank Layfield. The Inquiry was finished before Chernobyl, so the accident itself was *not* part of the public debate. But many people found it difficult to believe that the Inspector would not take the accident, its implications and its impact on public confidence in nuclear power into account in drafting his report and formulating his recommendations. It is notable that safety arguments about nuclear power took up much more time during the Inquiry than economic arguments.

The Layfield report was originally expected late summer in 1986, but publication was delayed for a number of months. The delay was never formally explained, but some critics insisted that it was to give the Government and the nuclear industry breathing space after Chernobyl, within which anti-nuclear fears and feelings might die down.

The report was eventually published on 26 January 1987, and it gave the go-ahead for building the PWR on the Suffolk coast. Layfield gave the nuclear industry the benefit of the doubt on safety issues and approved the proposal on grounds of national need. The Inspector had many reservations about the CEGB's economic case for the plant, but he still saw it as the least-cost choice for new generating capacity. He also recognized the dangers of low-level radiation and expressed reservations on the abilities of the emergency planning authorities to cope with a serious nuclear power accident, but conceded that the proposal appeared to be a "tolerable risk". He concluded that Sizewell B should be approved "in the national interest".

The Government was aware of the strength of anti-nuclear feelings around the country and conscious of the need to handle the Sizewell debate properly. It took the unusual step of debating the Layfield recommendations in Parliament. Some argue that Sizewell was a *fait accompli* once Layfield had come out in favour of it, because voting in the parliamentary debate was Party-based (and the Conservatives held the majority). The £1,600 million proposal appeared to offer cheap reliable energy supplies and promised to create around 10,000 jobs a year for seven years in the engineering and construction industries. The Government announced its approval on 12 March 1987. The opposition Labour Party reaffirmed its policy that it would cancel the PWR if re-elected to office at the next election (due that summer), even if construction work had already started.(Labour lost the election.)

The CEGB were relieved (but doubtless not surprised) to get Government approval for Sizewell, and they lost no time in pressing ahead with their plans. By late April 1987 they had invited tenders worth up to £500 million for civil engineering work associated with the project. They also unveiled plans for a further six PWRs in Britain by the turn of the century.

Public reactions

It is generally believed that nuclear power figures fairly low on people's lists of personal worries in Britain, but opinion polls have produced rather mixed results. For example, NOP Market Research Ltd published the results of a nationwide survey on 2 October 1986 which asked over 1,000 people "Which environmental problem (from a list given) do you think it is most important for the government to deal with?". Sixty percent selected nuclear power and nuclear waste, followed in rank order by use of chemicals in farming (54%), quality of drinking water (45%), acid rain (43%) then other environmental problems.

Different views emerged in a *Guardian* poll of 10-17 year olds around Britain, which asked them what problems (from a list given) worry them most facing Britain and facing the world. Nuclear power came a poor fourth (with 4%) at home, after unemployment (23%), nuclear weapons (19%) and crime and violence (8%). Nuclear power came eighth in the world (with 1%), after famine and poverty (24%), nuclear weapons (23%), wars (8%), politics (4%), violence and crime (3%), health (2%), and unemployment (2%).

Chernobyl *did* cause a surge of public anxiety over nuclear safety in Britain (see Chapter Six). But opinion poll results show that the accident seems to have had relatively little lasting effect on public attitudes in Britain. For example a BNFL survey published in December 1986 produced comforting results for the nuclear industry. It showed that public attitudes towards nuclear power had returned almost to pre-Chernobyl levels within six months. About half of the respondents felt that the amount of electricity produced by nuclear power should stay the same or be increased, 40% said it should be reduced, and the other 10% didn't know.

Nuclear industry's reactions

A new breath of openness and realism was added to the nuclear safety debate when Harold Bolter (Corporate Affairs Director of BNFL) told a nuclear conference at Lancaster University shortly after Chernobyl that the nuclear industry was "on probation". He added that "there are no short-cuts. Our self-scrutiny must be the most rigorous of all. The industry has got to deserve confidence before it can win or restore confidence. We must be extremely self-critical and accept constructive criticism from any source".

He was particularly conscious that the industry needs to sharpen up its distribution of factual material and information to the public. The House of Lords also picked out effective education about nuclear power as one of three responsibilities for industry to calm public fears, along with greater honesty in all branches of the civil nuclear industry and maintenance of its good track record. They also concluded that the nuclear industry needs to communicate in a language that the public can understand.

The nuclear industry (the CEGB and BNFL) responded by launching a costly image-building campaign designed to boost flagging public confidence in nuclear power. It included a nationwide publicity campaign, with full page spreads in leading national papers giving details of how nuclear power stations work, what safety features they have, how radiation affect health, and what are the inferred 'real' risks of nuclear power. CEGB produced a 14 minute colour video explaining how safe British nuclear reactors are, entitled *"It can't happen here"*, and distributed free copies to anyone who wrote in asking for one. Both CEGB and BNFL published glossy leaflets on nuclear power, and circulated them widely.

Confidence-building initiatives were also targeted on areas with existing nuclear installations. These included local media coverage, special open days at nuclear stations, and a greater profile of the stations and their personnel in local affairs (for example by mounting local displays on nuclear power and by sponsoring local sports events). BNFL invested heavily in new public exhibition facilities at Sellafield, hoping to attract large numbers of visitors and persuade them that nuclear energy is a safe form of technology.

This display of openness and apparent honesty within Britain's nuclear industry had a rather mixed reception. Some people welcomed it as long-overdue and sincere. Others dismissed it as cosmetic window-dressing which papered over the cracks in nuclear safety and simply drew attention to the industry's need to woo public support.

The House of Lords committee of inquiry on nuclear power concluded during the summer of 1986 that despite Chernobyl there is still a strong case for expanding nuclear power in Britain and other countries in the European Community. They believed that, although Chernobyl might bring a temporary pause in the expansion of the European nuclear programme, "the technological and economic case for developing nuclear power will prove *not* to have been weakened ... (and) ... nuclear power will provide, generally, the cheapest form of electricity, assisting the competitiveness of Europe's industries and raising living standards".

Most of Europe eventually decided to stick to their long-range plans to expand nuclear power. Some (like Finland and the Netherlands) decided to delay decisions on nuclear development. But others (like France and Britain) pressed ahead with a new determination and a renewed sense of national pride, having already established - at least to their own satisfaction - that "it *can't* happen here" (see Chapter Nine).

REACTIONS OUTSIDE EUROPE

USA

Chernobyl also came at an inopportune time in the United States, where - in the post-Three Mile Island era - there had been widespread public anxiety over plans for nuclear expansion, and many local battles over regulatory hearings and plant licensing. It was expected to galvanize public opposition to nuclear power in general, and strengthen the call for no more nuclear plants.

The USA had a large investment in nuclear capacity, and the stakes were high. Its 93 nuclear plants were providing over 77,000 megawatts (Table 10.1), around 15% of the nation's total electricity (Figure 10.2), and a further 26 plants with a combined capacity of nearly 30,000 megawatts were under construction.

The accident at Chernobyl woke up much anti-nuclear feeling in the USA which is reflected in poll results. For example, a survey by the national newspaper *USA Today*, soon after the accident, found that 58% of the people interviewed believe the kind of accident which happened at Chernobyl "can happen anywhere". A poll carried out for NBC News and the *Wall Street Journal* found 65% opposed to building more nuclear plants. Even greater anti-nuclear views were uncovered in a *Washington Post*-ABC poll (published on Saturday 24 May 1986), which showed a record 78% of people opposed to construction of more nuclear plants in the USA (compared with 65% in 1983 and 67% in 1985). That poll also found that 40% of people want existing plants phased out.

These public opinion surveys after Chernobyl reflect a widespread lack of faith in official statements on safety of nuclear energy, and a marked shift from the public confidence in nuclear energy being shown ten years earlier. The nuclear power industry was quick to point out that plant design and safety standards in the US are much tougher and better than in the Soviet Union (see Chapter Nine), but this did little to put people's minds at rest.

Many people in the United States believed that one immediate casualty of the accident was likely to be the Shoreham nuclear power station on Long Island in New York State. The plant had already been built by the Long Island Lighting Company and it was ready to operate by May 1986. But it had not yet been awarded its operating licence (by the Nuclear Regulatory Commission) because there was no suitable local evacuation plan for use in the event of an emergency. Basically there are only two ways off Long Island, both by bridge. The fear was that these would both very quickly be blocked with evacuation traffic, making it impossible for emergency services to move around. The speed and scale of evacuation needed at Chernobyl (see Chapter Four) probably added to the licensing authority's nervousness.

There was considerable concern in the nuclear industry and amongst politicians that growing public opposition to nuclear power might bring delays in the licensing of more than 20 nuclear plants expecting approval over the next few years. By mid-May 1986 Congress had postponed action (pending the outcome of a full evaluation of the Chernobyl accident), on reviewing the Price Anderson Nuclear Insurance Act and on reviewing procedures for licensing standardized reactor designs. Chernobyl had left its mark on the US public, and it was going to be an up-hill struggle to win back their support for and confidence in nuclear power.

Japan

Japan has few natural sources of energy, and has invested heavily in nuclear power. By May 1986 the country was getting nearly 25% of its electricity (Figure 10.2) from its 33 nuclear power stations (Table 10.1). A further 11 reactors were being built and 7 more were planned at that stage. Since the accident the authorities have revealed their intentions to build even more nuclear plants.

In July 1986 the nuclear sub-committee of the Japanese Advisory

Committee for Energy announced ambitious plans to more than double the nuclear capacity from nearly 24,000 megawatts to 62,000 megawatts by the year 2030. This will supply over half the country's predicted electricity needs, and require construction of two or three plants a year, bringing the total to 120. Some will be fast-breeder reactors.

A key element in the plan is to build a huge nuclear fuel recycling facility in Aomori Prefecture in the 1990s. This will be used for uranium fuel enrichment, fuel reprocessing and low level waste storage (see Figure 10.1).

The plans have aroused strong anti-nuclear feelings and attracted considerable opposition amongst the Japanese people. Rallies and demonstrations, aimed at stopping the nuclear expansion policies, have been held in major cities and at nuclear plants around the country. The country's Socialist and Communist parties have both called on the government to halt construction of new nuclear plants, and to carefully re-examine existing nuclear policies. To date the government has not taken their overtures seriously, and has not been deflected from its expansionist nuclear philosophy.

DEVELOPING COUNTRIES

Many developing countries have recognized the fundamental importance of having reliable energy supplies to meet the needs of industrial development, economic expansion and urban growth. But the introduction of nuclear power has been slower than some experts had expected. During 1986, for example, there were a total of 21 nuclear reactors in operation in developing countries, and a further 18 under construction.

Ten countries account for 63% of the total electricity production in the developing world, and eight of these have nuclear power programmes. But the distribution of nuclear capacity between countries is markedly uneven, with well over half of the reactors concentrated in two countries. India had six and was building four more at the time of Chernobyl, and the Republic of Korea had four and was building five more (Table 10.1).

Chernobyl is likely to have a minimal impact on nuclear power in developing countries, given their need for sustainable energy supplies and the lack of suitable, reliable and cost-effective alternatives. Electricity consumption is growing faster than primary energy consumption in the developing countries, and faster there than in the developed countries. The future is likely to see further expansion of nuclear power in those countries.

CONCLUSIONS

Many people have wondered what harm Chernobyl is likely to do to the nuclear industry world-wide. It is impossible to provide a firm answer, partly because the aftermath of the accident might be a long one and also because the 'harm' may not become apparent in the most obvious ways (such as closing down existing nuclear plants, as in Sweden).

There is little doubt that the accident focussed people's attentions on the inherent dangers of nuclear technology. Chernobyl itself showed just how serious and wide-ranging an accident in a nuclear reactor can be. The fact that an accident can occur and radiation leaks out was not entirely novel; Three Mile Island had convinced people around the world that such things *can* happen. What did shock many people was the vast quantity of radioactive material that leaked out, and the vast area over which it was distributed.

Those who previously thought that nuclear accidents can be limited to damage around the installation itself were forced to think again. Those who argued that there was no real problem because most countries have well designed emergency plans for coping with major radiological accidents had to change their tunes. Those who claimed that nuclear power was safe and reliable had to eat their words.

The period immediately post-Chernobyl was an era of reflection on the whole issue of nuclear power. Politicians, scientists, nuclear engineers, radiological protection people and the general public all had the need and the opportunity to take a long hard look at nuclear power. Some looked longer and harder than others!

Chernobyl cast a shadow over safety throughout the entire nuclear fuel cycle (see Figure 10.1). Many people, particularly in Europe, the United States and Japan, have expressed great concern over four particular areas of nuclear safety. First, and uppermost in people's minds, is the prospect of a reactor accident leading to release of radiation. Chernobyl confirmed their worst fears - it *can* happen, even if specialists insist that it *should not* have or *should not* happen again.

The second area of concern is over the possible effects of small quantities of radioactive effluents being discharged by nuclear plants into the environment during routine operations. This has been a particularly contentious issue at BNFL's Sellafield reprocessing plant in north-west England.

The third area is the whole issue of safe storage and ultimate disposal of radioactive wastes, especially those containing long-lived radio-isotopes. We have seen above the sort of public concern over storage which has been aroused in the United Kingdom, West Germany and Japan.

The fourth is over the transport of nuclear material from one part of the fuel cycle to another. For example, enriched fuel must be transported between fuel processing facilities and the nuclear plants, and spent fuel must be moved between plants and reprocessing facilities. All such movements pose risks of accidental leakage of radioactive materials, and many transport routes have no alternative but to pass through areas of high population.

Whilst the public are not always happy about nuclear power, the simple truth is that nuclear power appears to be here to stay. By the end of 1985 a total of 374 nuclear plants around the world were producing around 15% of the world's electricity. That same year the world's total installed nuclear capacity grew by 15% and 32 new reactors were brought into operation. There were a further 157 reactors under construction. The seven heads of state (from Canada, France, West Germany, Italy, Japan, UK and USA) who met in Tokyo in May 1986 (see Table 5.2) were united in their view that nuclear power is and will remain an increasingly widely used source of energy, provided that it is properly managed.

An editorial in *The Times* on 27 May 1987 sets the whole incident into context: "After Chernobyl, opponents of nuclear power can no longer be dismissed as green extremists or irrational non-scientists. The result could be and should be not an end to the development of nuclear power but a more sophisticated debate on its merits, leading to a safer, better understood and so ultimately more acceptable nuclear industry in both halves of Europe". The shame is that it has taken an accident so serious, widespread and long-lasting in impact to bring this about.

BIBLIOGRAPHY

Note: extensive use has also been made of coverage of Chernobyl in *The Times* between April 1986 and August 1988

Ahearne, J.F. (1987) Nuclear power after Chernobyl. *Science* 236; 673-9
Alexandropoulos, N.G. *et al* (1986) Chernobyl fallout on Ioannina, Greece. *Nature* 322; 779
Allman, W.F. (1986) Chernobyl: an overreaction? *Science* 86;
Anderson, A. (1986) Tokyo summit wants to know more. *Nature* 321; 186
Anderson, A. (1986) Making the most of Chernobyl. *Nature* 322; 673
Anon (1986) Chernobyl: Britain's N-power is safe. *BNFL News* (May); 2
Anon (1986) Cloud over Chernobyl. *Nature* 321; 2
Anon (1986) Nuclear cloud. *New Scientist* (1 May); 16
Anon (1986) The long shadow from Chernobyl. *Nature* 321; 99
Anon (1986) Anxiety about reactor accident subsides. *Nature* 321; 100
Anon (1986) Nuclear prognostications; estimating gravity of problems. *Nature* 321; 101
Anon (1986) Nuclear propaganda. *New Scientist* (8 May); 16
Anon (1986) Long haul ahead at Chernobyl. *Nature* 321; 183
Anon (1986) Chernobyl: intangible fallout from reactor. *Nature* 321; 186
Anon (1986) Ukraine may have contaminated soils for a decade. *New Scientist* (15 May); 25
Anon (1986) What future for nuclear power? *Nature* 321; 367
Anon (1986) Gorbachev blames Chernobyl on hydrogen explosion. *New Scientist* (22 May); 23
Anon (1986) Chernobyl: British embassies turn monitors. *Nature* 321; 458
Anon (1986) France discovers the nuclear scare. *New Scientist* (29 May); 22
Anon (1986) Nuclear industry is 'on probation'. *BNFL News* (June); 3
Anon (1986) Counting the score on transplants. *Nature* 321; 550
Anon (1986) Minister delayed ban on lamb slaughter. *New Scientist* (26 June); 25
Anon (1986) Half-way report on Chernobyl. *Nature* 322; 293
Anon (1986) Catechism for nuclear power. *Nature* 322; 393
Anon (1986) Chernobyl: experimental speculations. *Nature* 322; 399
Anon (1986) Nuclear regulation. *Nature* 322; 582
Anon (1986) Nuclear forecast. *Nature* 322; 586
Anon (1986) Chernobyl made plain. *Nature* 322; 672
Anon (1986) Timetable for a reactor disaster begins to emerge. *Nature* 322; 673
Anon (1986) Chernobyl report. *Nature* 323; 25-30
Anon (1986) Radiation, man and environment. *Nuclear India* 24; 3-5
Anon (1986) Soviet Union showed the world how to evacuate. *New Scientist* (4 September); 25

Anon (1986) Chernobyl 'has not hit case for A-power'. *BNFL News* (Sept); 5

Anon (1986) UKAEA faces up to Chernobyl. *Chemistry in Britain* 22; 983

Anon (1986) Russian reactor accident. *Atom* 356; 29

Anon (1986) Chernobyl accident. *Atom* 357; 16-7

Anon (1986) Chernobyl disaster. *Voice of Solidarity* 18-9; 28-35

Anon (1986) Chernobyl: clay cure for contaminated lamb. *Chemistry in Britain* (Sept); 779

Anon (1986) Chernobyl: lessons to be learned. *Chemistry in Britain* (June); 506-7

Anon (1986) There is a silent enemy lurking. *Time* (23 June); 19-21

Anon (1986) Radiation levels: WHO report on Chernobyl. *IAEA Bulletin* (Autumn); 27-9

Anon (1986) Post-accident review meeting. *IAEA Bulletin* (Autumn); 63-5

Anon (1986) Response to Chernobyl. *IAEA Bulletin* (Summer); 61-6

Anon (1987) Recalculating the cost of Chernobyl. *Science* 236; 658

Anon (1987) What are you worried about? *Guardian* (7 Jan); 11

Anon (1987) The Chernobyl disaster. *Which?* (April); 182-3

Aoyama, M. *et al* (1986) High level radioactive nuclides in Japan in May. *Nature* 321; 819-20

ApSimon, H. & J. Wilson (1986) Tracking the cloud from Chernobyl. *New Scientist* 111 (1517); 42-5

ApSimon, H.M., H.F. Macdonald & J.J.N. Wilson (1986) An initial assessment of the Chernobyl-4 reactor accident release source. *Journal of the Society for Radiological Protection* 6 (3); 109-19

Armstrong, J. (1985) The Sizewell Inquiry. *Journal of Planning and Environmental Law* (October); 686-9

Baes, C.F. *et al* (1986) Long term environmental problems of radioactively contaminated land. *Environment International* 12; 545-53

Banaschik, M.V. & K.H. Berg (1986) Federal Republic of Germany: steps for safety. *IAEA Bulletin* (Autumn); 35-6

Baransky, N.N. (1956) *Economic geography of the USSR.* Foreign Language Publishing House, Moscow

Beardsley, T. (1986) US analysis incomplete. *Nature* 321; 187

Beardsley, T. & V. Rich (1986) US physician tells heroic tale of Moscow. *Nature* 321; 369

Bennett, L.L. & R. Skjoeldebrand (1986) Worldwide nuclear power status and trends. *IAEA Bulletin* (Autumn); 40-5

Berman, M. (1986) The message of Chernobyl. *Resurgence* 119; 6-7

Berry, R.J. (1986) Living with radiation after Chernobyl. *The Lancet* (13 September); 609-10

Berry, R.J. (1987) Chernobyl; the anatomy of a disaster. *Cancer Topics* 6; 40-2

Black, D. (1987) New evidence on childhood leukaemia and nuclear establishments. *British Medical Journal* 294; 591-2

Blix, H. (1986) The post-Chernobyl outlook for nuclear power. *IAEA Bulletin* (Autumn); 9-12

Bondetti, E.A. & J.N. Brantley (1986) Characteristics of Chernobyl radioactivity in Tennessee. *Nature* 322; 313-4

British Nuclear Energy Society (1987) *Chernobyl; a technical appraisal.* BNES, Telford

Brown, M. & H.M. White (1987) The public's understanding of radiation and nuclear waste. *Journal of the Society for Radiological Protection* 7; 61-70

Bunyard, P. (1979) Living on a knife-edge - the aftermath of Harrisburg. *The Ecologist* 9; 97-102

Bureau Europeen des Unions de Consommateurs (1987) *Chernobyl - the aftermath.* BEUC, Brussels

Cambray, R.S. *et al* (1987) Observations on radioactivity from the Chernobyl accident. *Nuclear Energy* 26; 77-101

Camplin, W.C. *et al* (1986) Radioactivity in surface and coastal waters of the British Isles. Monitoring of fallout from the Chernobyl reactor accident. *Aquatic Environment Monitoring Report of the MAFF Directorate of Fisheries Research* 15; 1-49

Cassel, C.K. (1986) Political and medical lessons from Chernobyl. *Journal of the American Medical Association* 256; 630-1

Central Electricity Generating Board (1986) *Chernobyl: could it happen here?* CEGB Video, London

Chamberlain, A.C. (1987) Environmental impact of particles emitted from Windscale piles, 1954-57. *Science of the Total Environment* 63; 139-60

Chapman, N. & T. McEwan (1986) Geological solutions for nuclear wastes. *New Scientist* 111; 36-40

Clark, M.J. (1986) Fallout from Chernobyl. *Journal of the Society for Radiological Protection* 6; 157-61

Clark, W. (1986) Learning from the cloud. *Environment* 28; 1, 45

Clarke, R.H. (1986) Radiological aspects of Chernobyl in Western Europe. *Journal of the Society for Radiological Protection* 6; 139-41

Clarke, R.H. (1987) Reactor accidents in perspective. *British Journal of Radiology* 60; 1182-8

Codell, R.B. (1985) Potential contamination of surface water supplies by atmospheric releases from nuclear plants. *Health Physics* 49; 713-30

Collier, J.G. (1987) Chernobyl. *Chemical Engineering* 433; 5

Collier, J.G. & L.M. Davies (1986) *Chernobyl.* Central Electricity Generating Board

Collins, H. *et al* (1986) Information exchange and mutual emergency assistance. *IAEA Bulletin* (Autumn); 16-7

Council of Europe (1986) Simulating a major disaster in Europe; studying the lessons of Chernobyl. *Environmental Conservation* 13; 270-1

Courtis, M. (1987) After Chernobyl. *Municipal Review* 57; 244-5

Crick, M.J. & G.S. Linsley (1983) *An assessment of the radiological impact of the Windscale reactor fire, October 1957.* NRPB, Chilton

Csongor, E. *et al* (1986) Chernobyl fallout in Debrecen, Hungary. *Nature* 324; 40

Daglish, J. & J. Gittus (1986) Chernobyl and after. *Atom* 360; 6-8
Dempster, C. (1987) Co-operation to stop the nuclear clock. *The Times Higher Education Supplement* (16 Jan); 10
Department of Environment (1986) *Levels of radiactivity in the UK from the accident at Chernobyl.* HMSO, London (2pp)
Department of Environment (1986) *Levels of radioactivity in the UK from the accident at Chernobyl, USSR, on 26 April 1986; a compilation of the results of environmental measurement in the UK.* HMSO, London (176pp)
Devell, L. *et al* (1986) Initial observations of fall-out from the reactor accident at Chernobyl. *Nature* 321; 192-3 (corrigendum in *Nature* 321; 643)
Dobrev, B. & L. Spasov (1986) Experience and prospects in Bulgaria. *IAEA Bulletin* (Autumn); 55-6
Dotto, L. (1986) *Planet earth in jeopardy: environmental consequences of nuclear war.* Wiley, Chichester
Dunster, J. (1986) How we reacted to Chernobyl. *New Scientist* (5 June); 62-3

Edmundson, B. (1987) United Kingdom nuclear reactor design and operation. *British Journal of Radiology* 60; 1174-7
Edwards, M. (1987) Chernobyl - one year after. *National Geographic* 171; 632-53
Edwards, R. (1986) Chernobyl fall-out. *Sanity* (June); 4-5
Ehrlich, P.R. *et al* (1977) *Ecoscience.* Freeman, San Francisco
Eisenbud, E. (1973) *Environmental radioactivity.* Academic Press, London
Eisenbud, E. (1979) *Environment, technology and health.* Macmillan, New York
Etemad, S. (1987) Taking stock after Chernobyl. *Forum* (May); 20-1

Fernie, J. & S. Openshaw (1986) Who wants nuclear waste? *Geographical Magazine* 58; 63-8
Flavin, C. (1987) *Reassessing nuclear power - the fallout from Chernobyl.* Worldwatch Institute, Washington DC
Forastiere, F. *et al* (1985) Lung cancer and natural radiation in an Italian province. *Science of the Total Environment* 45; 519-26
Fowler, S.W. *et al* (1987) Rapid removal of Chernobyl fallout from Mediterranean surface waters by biological activity. *Nature* 329; 56-8
Fry, F.A. (1987) The Chernobyl reactor accident; the impact on the United Kingdom. *British Journal of Radiology* 60; 1147-58
Fry, F.A., R.H. Clarke & M.C. O'Riordan (1986) Early estimates of UK radiation doses from the Chernobyl reactor. *Nature* 321; 193-5
Fulker, M.J. (1987) Aspects of environmental monitoring by British Nuclear Fuels plc following the Chernobyl reactor accident. *Journal of Environmental Radioactivity* 5; 235-44
Funtowicz, S.O. & J.R. Ravetz (1986) Chernobyl's message. *Nature* (12 June); 25

Gale, R.P. & T. Hauser (1988) *Chernobyl; the final warning.* Hamish Hamilton, London

Geiger, H.J. (1986) The accident at Chernobyl and the medical responses. *Journal of the American Medical Association* 256; 609-12

Gelder, R. *et al* (1985) Consequences of UK transport accidents incidents involving radioactive materials, 1964-83. *Radiological Protection Bulletin* 67; 5-6

Gittus, J.H. *et al* (1987) *The Chernobyl accident and its consequences.* UKAEA. Oxford

Green, P. & P. Bailey (1987) *Fallout over Chernobyl.* Friends of the Earth, London

Greene, O. (1982) *London after the bomb. What a nuclear attack really means.* Oxford University Press, Oxford

Greenhalgh, G. (1986) The Soviet drive to nuclear power. *New Scientist* 110; 18

Greenwald, J. (1986) Deadly meltdown: a nuclear accident at Chernobyl. *Time* 127 (19); 6-15

Greenwald, J. (1986) More fallout from Chernobyl. *Time* (19 May); 6-8

Greenwald, J. (1986) Gorbachev goes on the offensive. *Time* (26 May); 10-11

Gubaryev, V. (1987) *Sarcophagus; a tragedy.* Penguin, Harmondsworth

Hamilton, E.I., B. Zon & R.J. Clifton (1986) The Chernobyl accident; radio-nuclide fallout in south west England. *Science of the Total Environment* 57; 231-51

Hamilton, M. (1986) Chernobyl; a cloud over the media. *Occupational Safety and Health* 16; 18-20

Hamilton, M. (1986) Chernobyl; would we get it right? *Occupational Safety and Health* 16; 21-3

Hamman, H. & S. Parrott (1987) *Mayday at Chernobyl.* New English Library, London

Hawkes, N. *et al* (1986) *The worst accident in the world. Chernobyl: the end of the nuclear dream.* Pan, London

Haynes, V. & M. Bojkun (1988) *The Chernobyl disaster.* Hogarth, London

Haywood, J.K. (editor) (1986) *Chernobyl; response of medical physics departments in the United Kingdom.* Institute of Physical Sciences in Medicine, London

Heath, M. (1985) Deep digging for nuclear waste disposal. *New Scientist* 108; 30-2

Hill, C.R. *et al* (1986) Iodine-131 in human thyroids in Britain following Chernobyl. *Nature* 321; 655-6

Hohenemser, C. *et al* (1986) Agricultural impact of Chernobyl; a warning. *Nature* 321; 817

Hohenemser, C. *et al* (1986) Chernobyl - an early report. *Environment* 28 (5); 6-13, 30-43

Holliday, B. *et al* (1986) Monitoring Minsk and Kiev students after Chernobyl. *Nature* 321; 820-1

Horrill, A.D. (1986) Radioactivity in terrestrial ecosystems. 42-47 in P. Ineson (editor) *Pollution in Cumbria.* Institute of Terrestrial Ecology, Huntingdon

Howells, K. (1986) Opposing nuclear power. *Sanity* (February); 14-7

Huda, W. (1986) Medical consequences of Chernobyl. *Journal of Ukrainian Studies* 20.

Hughes, J.S. (1985) The radiation exposure of the UK population. *Radiological Protection Bulletin* 62; 8-13

Hughes, J.S. & G.C. Roberts (1984) *The radiation exposure of the UK population; 1984 review.* National Radiological Protection Board, Chilton

Independent Advisory Group (1984) *Investigation of the possible increased incidence of cancer in West Cumbria.* HMSO, London

Ingham, J. (1986) Sellafield: best in the world. *BNFL News* (June); 1

International Atomic Energy Agency (1979) *Radiation - a fact of life.* IAEA, Vienna

International Atomic Energy Agency (1987) *One year after Chernobyl; the IAEA's actions and programmes in nuclear safety.* IAEA, Vienna

Jackman, B. (1986) The last round-up? *The Sunday Times Magazine* 30 Nov; 59-63

Jackson, D. *et al* (1987) Environmental monitoring in the vicinity of Sellafield following deposition of radioactivity from the Chernobyl accident. *Journal of the Society for Radiological Protection* 7; 75-87

Jaworowski, Z. (1986) Poland: the first four weeks. *IAEA Bulletin* (Autumn); 33-4

Jeffrey, J. (1986) The unique dangers of nuclear power; an overview. *The Ecologist* 4;5

Jensen, M. & J.C. Lindhe (1986) Sweden: monitoring the fallout. *IAEA Bulletin* (Autumn); 30-2

Jerome, F. (1986) Media Resource Service: getting scientists and the media together. *Impact of Science on Society* 144; 373-8

Johnston, K. (1987) United Kingdom upland grazing still contaminated. *Nature* 326; 821

Johnston, K. (1987) British sheep still contaminated by Chernobyl fallout. *Nature* 328; 661

Johnstone, B. (1987) Layfield gives approval to Sizewell B. *Nature* (19 Jan);

Jones, G.D. *et al* (1986) Observations of 110m-Ag in Chernobyl fallout. *Nature* 322; 313

Jones, R.R. (1986) Cancer risk assessments in light of Chernobyl. *Nature* 323; 585-6

Jost, D.T. *et al* (1986) Chernobyl fallout in size fractionated aerosol. *Nature* 324;22-3

Kelly, P. (1986) How the USSR broke into the nuclear club. *New Scientist* (8 May); 32-5

Kemp, R. *et al* (1986) Environmental politics in the 1980s: the public examination of radioactive waste disposal. *Policy and Politics* 14; 8

Kereiakes, J.G. *et al* (1986) The reactor accident at Chernobyl; a medical practitioners' perspective. *Seminars in Nuclear Medicine* 16; 224-30

Ketchum, L.E. (1987) Lessons from Chernobyl; Society of Nuclear Medicine members try to decontaminate world threatened by fallout. *Journal of Nuclear Medicine* 28; 933-42

Ketchum, L.E. (1987) Lessons of Chernobyl; health consequences of radiation released and hysteria unleashed. *Journal of Nuclear Medicine* 28; 413-22

Kingman, S. (1986) A lot of fuss about a few millisieverts. *New Scientist* (15 May); 26

Kolata, G. (1986) The UCLA-Occidental-Gorbachev connection. *Science* 233; 19-21

Kolata, G. (1986) Soviets presented plans for Chernobyl study. *Science* 233; 513-4

Levi, B.G. (1986) Cause and impact of Chernobyl accident still hazy. *Physics Today* 39; 17-21

Lewis, H.W. (1986) The accident at the Chernobyl nuclear power plant and its consequences. *Environment* 28; 25-7

Loprieno, N. (1986) Radiation knows no frontiers. *European Environment Review* 1; 2-9

Macgill, S.M. (1987) *The politics of anxiety; Sellafield's cancer link controversy.* Pion, London

Mackay, L. & M. Thompson (1988) *Something in the wind; politics after Chernobyl.* Pluto, London

Mackenzie, D. (1986) Germany fights over the future of nuclear power. *New Scientist* (29 May); 22

Macleod, G.K. & W.R. Hendee (1986) Radiation accidents and the role of the physician; a post-Chernobyl perspective. *Journal of the American Medical Association* 256; 632-4

Marples, D.R. (1986) Chernobyl and Ukraine. *Problems in Communism* (Nov-Dec); 25

Marples, D.R. (1986) *Chernobyl and nuclear power in the USSR.* Macmillan and Canadian Institute of Ukrainian Studies, University of Alberta

Marshall, E. (1986) Reactor explodes amid Soviet silence. *Science* 232; 814-5

Marshall, E. (1986) The lessons from Chernobyl. *Science* 233; 1375-6

Marshall, Lord (1986) *The Chernobyl accident.* CEGB, London

Marshall, Lord (1986) United Kingdom: Chernobyl - the aftermath. *IAEA Bulletin* (Autumn); 36-8

Marwick, C. (1986) Physicians' reaction to Chernobyl explosion; lessons in radiation - and cooperation. *Journal of the American Medical Association* 256; 559-65

Mason, I. (1986) Introducing the national cabbage monitoring network. *New Scientist* (22 May); 23

Matthews, R. (1979) *Nuclear power and safety.* CEGB, London

Medvedev, Z. (1976) Two decades of dissidence. *New Scientist* 72; 264-7

Medvedev, Z. (1979) *Nuclear disaster in the Urals.* Angus and Robertson, London

Medvedev, Z. (1986) Soviet power failures. *Marxism Today* (June); 10-15

Medvedev, Z. & A. Roberts (1977) *Hazards of nuclear power.* Spokesman, Nottingham

Miller, G.T. (1979) *Living in the environment.* Wadsworth, Belmont California

Milne, R. (1987) Inquiry backs PWR for Sizewell. *New Scientist* 113; 22-3

Ministry of Agriculture, Fisheries and Food (1987) *Radionuclide levels in food, animals and agricultural products; post Chernobyl monitoring in England and Wales.* MAFF, London

Mitchell, N.T. *et al* (1986) The Chernobyl reactor accident and the aquatic environment of the United Kingdom; a fisheries viewpoint. *Journal of the Society for Radiological Protection* 6; 167-72

Morrey, M. *et al* (1987) *A preliminary assessment of the radiological impact of the Chernobyl reactor accident on the population of the European Community.* NRPB, Oxford.

Mossmay, K.L., D.L. Thomas & A. Dritschilo (1986) Environmental radiation and cancer. *Journal of Environmental Science and Health* C4; 119-61

Mould, R.F. (1987) After Chernobyl. *British Institute of Radiology Bulletin* (June); B29-B34

Mould, R.F. (1988) *Chernobyl; the real story.* Pergamon, Oxford

Moundrey, P.J. *et al* (1986) Transfer of radioiodine to milk and its inhibition. *Nature* 322; 600

Nair, S. & P.J. Darley (1986) A preliminary assessment of individual doses in the environs of Berkeley, Gloucestershire, following the Chernobyl nuclear reactor accident. *Journal of the Society for Radiological Protection* 6 (3); 101-18

Nair, S. *et al* (1986) Nuclear power and the terrestrial environment; the transport of radioactivity through foodchains to man. *CEGB Research* 19; 3-16

National Academy of Sciences (1980) *The effects on populations of exposure to low levels of ionizing radiation.* National Academy of Sciences, Washington DC

Nechaev, A., V. Onufriev & K.T. Thomas (1986) Long-term storage and disposal of spent fuel. *IAEA Bulletin* (Spring); 15-20

Needham, G. (1986) Ackworth takes charge. *New Scientist* (12 June); 57-8

Neffe, J. (1986) Germany and Chernobyl - end of the nuclear programme? *Nature* 321; 640

Nehot, J.C. (1987) Medical basis for the establishment of intervention levels. *British Journal of Radiology* 60; 1163-69

Nero, A.V. *et al* (1986) Distribution of airborne radon-222 concentrations in US homes. *Science* 234; 992-7

Nishizawa, K. *et al* (1986) Iodine-131 in milk and rain after Chernobyl. *Nature* 324; 308

NOP Market Research Ltd (1986) Acid rain. *Political, Social, Economic Review* 61; 9-14

Norman, C. (1986) Hazy picture of Chernobyl emerging. *Science* 232; 1321-3

Norman, C. (1986) Chernobyl-type accident deemed unlikely at Hanford plant. *Science* 233; 837-8

Norman, C. (1986) Chernobyl: errors and design flaws. *Science* 233; 1029-30

Norman, C. & D. Dickson (1986) The aftermath of Chernobyl. *Science* 233; 1141-3

Nuclear Energy Agency (1986) *Nuclear reactor accidents; source terms.* Nuclear Energy Agency, Paris

Nuclear Energy Agency (1987) *Chernobyl and the safety of nuclear energy reactors in OECD countries.* HMSO, London

Openshaw, S. (1986) *Nuclear power: siting and safety.* Routledge, London

Organization for Economic Cooperation and Development (1987) *Chernobyl and the safety of nuclear reactors in the OECD countries.* OECD, Paris

Patterson, W.C. (1986) Chernobyl; worst but not first. *Bulletin of the Atomic Scientists* (August);

Patterson, W.C. (1986) Chernobyl; the official story. *Bulletin of the Atomic Scientists* (November);

Persson, C., H. Rodhe and L.E. de Greer (1987) The Chernobyl accident; a meteorological analysis of how radionuclides reached and were deposited in Sweden. *Ambio* 16; 20-31

Petrosyants, A. (1986) The Soviet Union and the development of nuclear power. *IAEA Bulletin* (Autumn); 5-8

Pohl, F. (1987) *Chernobyl; a novel.* Bantam, London

Pollock, C. (1986) The closing act; decommissioning nuclear power plants. *Environment* 28; 10-15, 33-36

Porritt, J. (1986) Chernobyl. *Resurgence* 117; 5

Porter, D. (1986) The Porterfolio. *Third Way* 9; 31

Pourchet, M. *et al* (1986) The northerly extent of Chernobyl contamination. *Nature* 323; 676

Preston, A. (1977) The study and control of environmental radioactivity, and its relevance to the control of other environmental contaminant. *Atomic Energy Review* 15; 371-405

Priest, J. (1973) *Problems of our physical environment.* Addison-Wesley, Reading, Massachusetts

Pringle, D.M. *et al* (1986) Gamma-ray spectrum at Chernobyl fallout. *Nature* 321; 569

Pryde, P.R. (1978) Nuclear energy development in the Soviet Union. *Soviet Geography* 19; 75-83

Pyatt, F.B. (1987) After Chernobyl; some data on radioactivity in Scandinavia. *International Journal of Environmental Studies* 29; 197-9

Quick, J. (1987) What have we learnt from Chernobyl? *Atom* 366; 1

Ratnieks, H. (1979) Soviet nuclear energy without restraint. *Geographical Magazine* (October); 1-3

Ray, P. (1986) A dearth of dollars. *The Traveller* 17; 20-1

Rich, V. (1986) Soviet nuclear reactors: ambitious plans with red tape. *Nature* 321; 100-1

Rich, V. (1986) Where does the blame lie? *Nature* 321; 187

Rich, V. (1986) Fusion may be a better way. *Nature* 321; 550

Rich, V. (1986) Chernobyl: now for the long haul. *Nature* 321; 641

Rich, V. (1986) Chernobyl: eastern bloc reactions. *Nature* 321; 805

Rich, V. (1986) Chernobyl: inquest continues on nuclear power. *Nature* 322; 203

Rich, V. (1986) Chernobyl accident is blamed on human error. *Nature* 322; 295

Rich, V. (1986) Chernobyl accident: reactor design not perfect. *Nature* 322; 588

Rich, V. (1986) Chernobyl accident: fallout pattern puzzles Poles. *Nature* 322; 765

Rich, V. (1986) Chernobyl: the international dimension. *The World Today* 42 (11);

Rich, V. (1987) More compensation in Finland for nuclear accident victims. *Nature* 325; 654

Rosen, M. (1986) New directions in nuclear safety. *IAEA Bulletin* (Autumn); 13-5

Royal Commission on Environmental Pollution (1976) *Sixth report: nuclear power and the environment.* HMSO, London

Rudolph, R. & S. Ridley (1986) Chernobyl's challenge to anti-nuclear activism. *Radical America* 20; 7-21

Rusche, B.C. (1986) Managing high-level waste in the USA. *IAEA Bulletin* (Spring); 48-52

Ryder, E.A. (1987) Regulation of nuclear power in the UK after Chernobyl. *Atom* 366; 8-12

Saire, D.E. (1986) World status of radioactive waste management. *IAEA Bulletin* (Spring); 5-9

Salo, A. (1986) Information exchange after Chernonyl. *IAEA Bulletin* (Autumn); 18-22

Sands, P. (1988) *Chernobyl; law and communication.* Grotius, Cambridge

Scarlott, J. (1986) Chernobyl, USA. *Radical America* 20; 12-5

Scheer, J. (1987) How many Chernobyl fatalities? *Nature* 326; 449

Schelenz, R. & A. Abdel-Rassoul (1986) Report from Sieberdorf: post-accident radiological measurements. *IAEA Bulletin* (Autumn); 23-6

Schreiber, T. (1986) Tchernobyl et les medias en Europe de l'Est. *Politique Etrangere* 3;

Schultz, V. & F.W. Whicker (1985) Ionizing radiation and nuclear war: review of deliberations on ecological impacts. *Critical Reviews in Environmental Control* 15; 417-27

Searle, C. (1987) Green and poisoned land. *Sanity* 4; 26-8

Serrill, M.S. (1986) Anatomy of a catastrophe. *Time* 128; 6-11

Shabad, T. (1986) Geographic aspects of the Chernobyl nuclear accident. *Soviet Geography* 7; 504-28

Sjoblom, K. & A. Toivola (1986) Operations experience in Finland. *IAEA Bulletin* (Autumn); 57-9

Smith, F.B. & M.J. Clark (1986) Deposition of radionuclides from the Chernobyl cloud. *Nature* 322; 690-1

Smith, G. (1987) Are the horsemen in the saddle? *Buzz* (January); 22-5

Snell, V.G. & J.Q. Howieson (1986) *Chernobyl; a Canadian perspective.* Atomic Energy of Canada.

Steinhausler, F. *et al* (1985) The main inconsequences in the present radiological protection concept for the general population. *Health Physics* 49; 1229-38

Swinbanks, D. (1986) No nuclear doubts for Japan. *Nature* 322; 399

Tatu, M. (1986) Un test pour le systeme. *Politique Etrangere* 3;

Taylor, P. (1981) *The Windscale fire, 1957.* Political Ecology Research Group, Oxford

Taylor, P. (1986) Chernobyl: the long term consequences. *New Scientist* (15 May); 24

Thies, J. (1986) Les consequences de Tchernobyl: un atout pour les relations Est -Ouest. *Politique Etrangere* 3;

Torrey, L. (1979) The accident at Three Mile Island. *New Scientist* (8 November); 424-8

Thomas, A.J. & J.M. Martin (1986) First assessment of Chernobyl radioactive plume over Paris. *Nature* 321; 817-9

Thornton, J. (1986) Chernobyl and Soviet energy. *Problems of Communism* (Nov-Dec); 8

Trabalka, JR. *et al* (1980) Analysis of the 1957-1958 Soviet nuclear accident. *Science* 209; 345-53

Trichopoulos, D. *et al* (1987) The victims of Chernobyl in Greece; induced abortions after the accident. *British Medical Journal* 295; 1100

Tyler Miller, G. (1979) *Living in the environment.* Wadsworth, Belmont

USSR State Committee on the Utilization of Atomic Energy (1986) *The accident at the Chernobyl nuclear power plant and its consequences.* Report to IAEA Experts' Meeting, Vienna (August)

Upton, A. (1982) The biological effects of low-level ionizing radiation. *Scientific American* 126; 29

Van Der Veen, J. *et al* (1986) Core fragments in Chernobyl fallout. *Nature* 323; 399-400

Varley, J. (1986) Chernobyl prepares to start up. *Nuclear Engineering International* 31; 7

Vendryes, G. (1986) Observations from France. *IAEA Bulletin* (Autumn); 52-4

Walker, P. (1986) Nuclear power - our only chance. *BNFL News* (July); 2-3

Walske, C. (1986) United States: lessons of Chernobyl. *IAEA Bulletin* (Autumn); 38-9

Webb, G.A.M. *et al* (1986) Radiation levels in Eastern Europe. *Nature* 321; 821-2

Webb, R.E. (1986) Chernobyl; what could have happened? *The Ecologist* 16;

Webster, D. (1986) How ministers misled Britain about Chernobyl. *New Scientist* 112; 43-6

Weinberg, S. (1986) Armand Hammer's unique diplomacy. *Bulletin of the Atomic Scientists* 43; 50-2

White, D. (1987) Sizewell; the votes in nukes. *New Society* (30 Jan); 16-7

White, P. (1986) For the record ... *IAEA Bulletin* (Spring); 69

Wild, G. (1986) L'economie sovietique d'apres Tchernobyl. *Politique Etrangere* 3;

Wilkie, T. (1986a) The unanswered questions of Chernobyl. *New Scientist* (15 May); 23

Wilkie, T. (1986b) Radiation? It's as plain as the nose on your face. *New Scientist* (12 June); 29

Wilkie, T. & R. Milne (1986) The world's worst nuclear accident. *New Scientist* (1 May); 17-9

Wilkie, T. & R. Milne (1986) Chernobyl: sorting fact from fiction. *New Scientist* 110; 17-20

Wilson, R. (1986) Chernobyl; assessing the accident. *Issues in Science and Technology* (Fall);

Woodwell, G. (1986) Chernobyl; a technology that failed. *Issues in Science and Technology* (Fall);

World Health Organization (1986) *Chernobyl reactor accident: report of a consultation, 6 May 1986, provisional.* WHO, Copenhagen (38pp)

Wright, J.K. (1987) Emergency planning. *British Journal of Radiology* 60; 1177-80

Wright, P. & T. Prentice (1986) Inside Chernobyl's cloud. *The Times* (7 May); 11

Wynne, B. (1978) The politics of nuclear safety. *New Scientist* 77; 208-11

Index